Aerospace Engineering:
From the Ground Up

Aerospace Engineering: From the Ground Up

Ben Senson
James Madison Memorial High School, Madison, Wisconsin

Jasen Ritter
Vandegrift High School, Austin, Texas

DELMAR
CENGAGE Learning™

Australia • Brazil • Japan • Korea • Mexico • Singapore • Spain • United Kingdom • United States

DELMAR
CENGAGE Learning™

Aerospace Engineering: From the Ground Up
Ben Senson, Jasen Ritter

Vice President, Career, Education, and Training Editorial: Dave Garza

Director of Learning Solutions: Sandy Clark

Senior Acquisitions Editor: James DeVoe

Managing Editor: Larry Main

Product Manager: Mary Clyne

Development: iD8 Publishing Services, Inc.

Editorial Assistant: Cris Savino

Vice President, Career, Education, and Training Marketing: Jennifer McAvey

Marketing Director: Deborah Yarnell

Marketing Manager: Katherine Hall

Production Director: Wendy Troeger

Production Manager: Mark Bernard

Content Project Manager: Mike Tubbert

Art Director: Casey Kirchmayer

Library of Congress Control Number: 2010936367

ISBN-13: 978-1-4354-4753-0

ISBN-10: 1-4354-4753-0

Delmar
5 Maxwell Drive
Clifton Park, NY 12065-2919
USA

Cengage Learning is a leading provider of customized learning solutions with office locations around the globe, including Singapore, the United Kingdom, Australia, Mexico, Brazil and Japan. Locate your local office at: **international.cengage.com/region**

Cengage Learning products are represented in Canada by Nelson Education, Ltd.

For your lifelong learning solutions, visit **delmar.cengage.com**

Visit our corporate website at **cengage.com**.

Notice to the Reader

Printed in the United States of America
1 2 3 4 5 6 7 15 14 13 12 11

BRIEF CONTENTS

TABLE OF CONTENTS

If you ask children what interests them the most, they will rattle off a list that includes dinosaurs, aliens, pirates, and space. Ask them what they dream about being, and very frequently you will hear about flying or space travel. Children dream about the freedom and wonder of the aerospace world. Indeed, flight remains one of the most awe-inspiring endeavors undertaken by humankind. Today we stand poised for the grandest adventure of all as we consider interplanetary travel, the colonization of new worlds, and the regularly scheduled travel of citizens into outer space. There are even space hotels being planned and tested. This is an exciting time to be alive!

And yet, very little material exists to guide students to an understanding of how humans conquered the once impossible challenge of flight. *Aerospace Engineering: From the Ground Up* was written to fill that gap. It was not written specifically to support pilots in training, nor is it intended for a very narrow audience. This text was written for anyone considering a personal involvement with flight or space exploration. It is appropriate for high school students in aviation classes, pilots in training, as well as students considering collegiate studies of aerospace related subjects such as aerospace engineering, air traffic control, and aviation management, among many others. You will not need any advanced mathematics to understand the concepts discussed in this book. You need only the basics of algebra and geometry.

AEROSPACE ENGINEERING AND PROJECT LEAD THE WAY, INC.

This text resulted from a partnership forged between Delmar Cengage Learning and Project Lead The Way, Inc. in February 2006. As a non-profit foundation that develops curriculum for engineering, Project Lead The Way, Inc. provides students with the rigorous, relevant, reality-based knowledge they need to pursue engineering or engineering technology programs in college.

The Project Lead The Way® curriculum developers strive to make math and science relevant for students by building hands-on, real-world projects in each course. To support Project Lead The Way's® curriculum goals, and to support all teachers who want to develop project/problem-based programs in engineering and engineering technology, Delmar Cengage Learning is developing a complete series of texts to complement all of Project Lead the Way's® nine courses:

Gateway To Technology
Introduction To Engineering Design
Principles Of Engineering
Digital Electronics
Aerospace Engineering

Biotechnical Engineering
Civil Engineering and Architecture
Computer Integrated Manufacturing
Engineering Design and Development

To learn more about Project Lead The Way's® ongoing initiatives in middle school and high school, please visit www.pltw.org.

HOW THIS TEXT WAS DEVELOPED

Aerospace engineering is a fascinating subject that illuminates the interface between history, science, engineering, and the very human characteristics of creativity and perseverance. This textbook was written to provide the broadest possible overview of the subject from the perspective of the engineer. It is intended to provide the background and supporting information required to appreciate the aviation and space related achievements of the last few centuries. As such we sought to provide the historical context and stories that bring the facts, figures, and skills to life as well as the necessary material to become literate about what aerospace engineers do in their daily work.

Although professionals in the aerospace engineering field will find that this coverage is much broader than their daily experience, we felt that it was essential to expose the student to the broadest possible range of activities related to the subject of aerospace studies. Therefore, this textbook is intended for students who want to explore the possibilities of direct participation in the aerospace industry as hobbyists, technicians, professionals, or simply as informed knowledgeable citizens.

We hope that you find it an inspiration to learn more.

Organization

This textbook has been organized to allow the reader to master fundamental material which is then built upon layer by layer. We begin with an introduction to the aerospace field and history of flight. We will then guide readers through the fundamentals of aerodynamics, control systems and propulsion. Applying these sciences to the world of flight is essential for building an understanding of the 'how' of aerial vehicles.

Beyond the physics and mechanics of flight, we will also discuss principles of navigation and aerospace physiology. Understanding how human physiology places constraints on the performance of an aerospace vehicle gives us the basis upon which to consider the trend in modern aerospace engineering toward the UAV, or unstaffed aerial vehicle, and similar technologies for surface and space operations.

When you finish reading this text, you will be able to make an informed decision about your future in aerospace engineering. Whether you choose to be a pilot, air traffic controller, mechanic, ground operations technician, avionics technician, load master, mission specialist, launch controller, satellite controller, aerospace engineer, or one of the hundreds of other related careers, this textbook has been organized to give you a basic understanding of the field from the perspective of those who design and optimize the solutions to aerospace challenges.

Features

Teachers want an interactive text that keeps students interested in the story behind the aerospace innovations that have shaped our world. This text delivers that story with plentiful boxed articles and illustrations that highlight the aerospace story. Here are some examples of how this text is designed to keep students engaged in a journey through aerospace engineering.

▶ **Case Studies** let students explore how aerospace engineering evolved and where this exciting field may evolve into the future.

Case Study ⫸

RAPID TRANSPORTATION ACROSS INLAND SEAS

In the former Soviet Union, numerous aircraft have been specifically designed to "fly" in ground effect to rapidly deliver massive cargos across the Caspian Sea. These specialty aircraft, known as ekranoplans, never fly more than a few hundred feet above the tops of the waves yet are capable of very rapidly delivering large amounts of cargo very efficiently (see Figure 3-16). The ekranoplan offers the advantages of an airplane's rapid transport with the ship's fuel efficiency.

Ekranoplans, or ground effect aircraft, are only effective above very flat land or water as any significant variation in height tremendously increases the chances for a collision with the surface.

Two ekranoplans could be seen on Google Earth located at 42°52′54″N,

47°39′24″E and at 42°52′50″N, 47°39′57″E. A structure on a nearby beach may be a third disassembled ekranoplan.

Figure 3-16 *An ekranoplan aircraft designed specifically for flight within ground effect.*

© Cengage Learning 2012

▶ **Boxed Articles** highlight fun facts and points of interest as the aerospace engineering industry took off.

Point of Interest

How Gliders Fly

Gliders tap into external energy sources such as a tow plane, a truck to pull them to the mountain top (or your legs!), stretched bungee cords or winches on the ground, and rising air currents due to the warm air rising in a thermal or orographic lifting as air flows over rising ground. Once in flight, gliders descend through the air, but if you are skilled, you may be able to climb by flying into air that is rising as a mass faster than your rate of descent within the air mass.

Your Turn

Planetary Pencil Experiment

You can make an ellipse easily with a loop of string, two pushpins, and a pencil.

Materials
2 pushpins, pencil, loop of string, bulletin board or similar material

Steps
1. Push the pushpins into the bulletin board a few inches apart from each other.
2. Make a loop of string that is a little larger than the distance between the two pins. Put the loop of string around the two pins.
3. Put the pencil inside the loop of string and gently pull outward to tighten the loop. Move the pencil in a circular motion while keeping the string under the same tension.

For fun, try moving the foci closer together or farther apart. What do you get when both foci are in the same place (only one pin)? How could you calculate or predict the size of your ellipse?

▶ **Your Turn** activities reinforce text concepts with skill-building activities.

▶ **Key Terms** are defined throughout the text to help students develop a reliable lexicon for the study of engineering.

Rocket Propulsion: Solid or Liquid?

The tradeoffs between two designs can be dramatic. For example, we can use a rocket motor to create the thrust to maneuver a spacecraft. Using a solid rocket has the advantage of being ready to fire at a moment's notice because its fuel is stable and easy to store. Liquid-fueled rockets have the advantage of being able to be turned on and off repeatedly, but their fuel is more challenging to store. So which one should be used? Aerospace engineers know that both are useful and rely on their experiences with them in the past to help choose which is the best solution for the current mission design.

Aerodynamics:

the study of the effect of air flowing over and interacting with objects and the production of forces such as lift and drag.

▶ **STEM** connections show examples of how science and math principles are used to solve problems in engineering and technology.

▶ **Career Profiles** provide role models and inspiration for students to explore career pathways in aerospace engineering.

▶ **Grass Strip Adventures** are included where relevant to list further reading or resources for research and projects.

Want to see computational fluid dynamics in action? Check out two videos from PBS entitled *Battle of the X-Planes* and *21st Century Jet: The Building of the 777*. Both show airplanes being designed from concept to initial testing using computer modeling, simulation, and analysis as well as prototyping and flight testing.

Careers in Aerospace Engineering

READY FOR LAUNCH

NASA's space program is a risky operation, but it's a little less so thanks to Harmony Myers. Myers is a safety engineering section chief, supervising 12 other engineers who keep an eye out for bugs in the system. "We look for things that could cause harm to the Space Shuttle or the astronauts," she says. "We try to eliminate or minimize them so that it's a safer flight program."

On the Job

NASA launches the Space Shuttle four or five times a year, and Myers and her team are right there in the control room. She talks to NASA engineers about potential problems with the shuttle's hardware. If the launch doesn't meet requirements, she recommends that it be postponed.

Myers is also involved in the prelaunch stage. She examines the shuttle's booster segments, looking for parts that might fail. She also checks out the system that monitors leaks in the shuttle's hydrogen and oxygen lines.

"The job is very challenging," Myers says, "and it changes day to day. Knowing you're part of getting the shuttle off the ground is inspiring. Every day, you're just grateful you can contribute to something wonderful like that."

Inspirations

Myers always loved math, and she was always curious about the way machines worked.

"I liked to take things apart and put them back together, like telephones," she says. "My parents probably hated me." In high school, Myers had a chance to shadow a professional for a day as part of a career

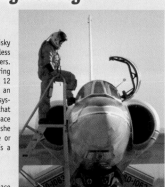

Pilot James Barrilleaux enters the cockpit of a NASA ER-2 aircraft. ER-2s are civilian versions of the military U-2S reconnaissance aircraft. As part of NASA's Airborne Science program, ER-2s can carry scientific payloads of up to 2,600 pounds to altitudes of about 70,000 feet to investigate such matters as earth resources, celestial phenomena, atmospheric chemistry and dynamics, and oceanic processes.

Source: NASA/Ames Research Center

program. "There was a list of professions, and I picked an engineer," she says.

She immediately knew that engineering was the career for her.

Education

Myers earned a Bachelor of Science degree in electrical engineering at the University of Central Florida and a master's degree in industrial engineering at the University of Miami.

"You learn a lot of math and science in college," Myers says, "but the main thing is to learn how to think analytically and logically through a problem."

As an undergraduate, Myers was inspired by the chance to work with Walt Disney World Ride and Show Engineering on a school project. She and a team of students were asked to work with a cable box used in Disney hotel rooms.

"It was really neat," Myers says. "We were given the box, and it was up to us to create an interface. We wired it to the thermostat so you could change the temperature in the room from your TV. It was a chance to brainstorm and figure out how to solve a problem on your own. That helped prepare us for the future of solving problems."

Advice to Students

Myers suggests that students talk to professional engineers about their jobs, just as she did. She also recommends taking as much math as possible, as early as possible, including calculus.

"As an engineer, you get to come up with solutions on your own to help better the environment, or to make things faster and quicker," Myers says. "And you can see your results. It's a very rewarding career."

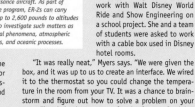

▶ **Summary:** at the end of each chapter provides a bulleted list of the major concepts presented throughout the chapter

Arrived at Destination

SUMMARY

- Milestones seem to be coming at exponential rates. For centuries, fire and rudimentary simple machines were state of the art technology. Over the span of a couple hundred of years, the steam engine transformed industry and transportation.

- The time span from the first flight of the Wright brothers to that of the Space Shuttle was not even 90 years. Technological innovation will continue to allow us to develop aircraft and spacecraft capable of performing feats previously considered impossible.

- In the modern classroom, most students carry a calculator with more computing power than the *Apollo* spacecraft took to the Moon. In the span of a decade, the U.S. space program progressed from barely putting a man in space to several missions on the Moon—a feat so monumental that a few people continue to claim it's a hoax.

- Often science fiction is the precursor to science fact. Much of the make-believe technology

of the 1960s and 1970s sci-fi shows, such as personal communicators, has become today's reality. Where space travel brought dreams of flying cars and Stanley Kubrick's space station of *2001: A Space Odyssey*, of exploration and colonization of space, our present technology creates possibilities that stir the imagination. Humans are driven to explore the unknown and to travel to wherever nature allows.

- The technology and the aircraft developed along the way—miniaturization of electronics, predictable flight characteristics, sophisticated propulsion systems, autonomous guidance systems, even Velcro—have become necessities of our daily life. Tsiolkovsky may have been correct in thinking humanity is like a bird in a nest. We are capable of overcoming great challenges when bright minds work together. If we are to truly unlock our creative potential and prevent stagnation as a species, we must leave the nest and further explore our universe.

BRING IT HOME

▶ **Bring It Home:** activities are provided at the end of each chapter. The activities progress in rigor from simple, directed exercises and problems to more open-ended projects.

1. Briefly describe the major innovation(s) developed by each of the following inventors:
   - Aristotle
   - Montgolfier brothers
   - Robert brothers
   - Daniel Bernoulli
   - George Cayley
   - Otto Lilienthal
   - Octave Chanute
   - Orville and Wilbur Wright
   - Samuel Langley
   - Frank Whittle and Hans von Ohain
   - Konstantin Tsiolkovsky
   - Robert Goddard
   - Wernher von Braun

2. Create a poster board that summarizes one aerospace innovation. Describe the inventor, the innovation, and how it improved the airplanes or spacecraft that followed.

EXTRA MILE

▶ **Extra Mile:** at the end of each chapter provides extended learning opportunities for students who want an additional challenge.

1. Construct a model kit of an airplane or spacecraft that interests you.
2. Construct a flying model airplane or rocket. (Safety guidelines for actual flight can be found at the following websites. You should always follow safety procedures when operating model aircraft and rockets!)
   - National Association of Rocketry (www.nar.org/)

   - Academy of Model Aeronautics (www.modelaircraft.org)
3. Build a full-scale mockup of a *Mercury*, *Gemini*, or *Apollo* space capsule out of cardboard.
4. Design a history of aviation mini-museum in a display case.

Supplements

A complete supplements package accompanies this text to help instructors implement 21st-century strategies for teaching aerospace engineering:

▶ A **Student Workbook** reinforces text concepts with practice exercises and hands-on activities.

▶ An **Instructor's e-resource** includes solutions to text and workbook problems, instructional outlines and helpful teaching hints, a STEM mapping guide, PowerPoint presentations, and computerized testing options.

HOW ENGINEERING DESIGN SUPPORTS STEM EDUCATION

Math and science are the languages we use to communicate ideas about engineering and technology. It would be difficult to find even a single paragraph in this text that does not discuss Science, Technology, Engineering, or Mathematics. The authors of this text have taken the extra step of showing the links that bind math and science to engineering and technology. The STEM icon shown here highlights passages throughout this text that explains how engineers use math and science principles to understand aerospace engineering. In addition, the Instructor's e-resource contains a STEM mapping guide to this career cluster

ACKNOWLEDGMENTS

This work, like all other written works, is the product of many minds and hands. Although we have poured our minds, bodies, and spirits into this work in order to make it useful to students and teachers interested in learning about the field of aerospace engineering, it could not have been completed without the supporting work of many individuals and organizations.

We would like to thank our families for granting us the time to hide away from the rest of the world to wrestle with the demons of confusion and rambling speech. We could not have succeeded in this task without the support of Lisa, Aaron, Braxton, MaiaLynn, Casey Jo, and Levi.

Thank you to the Cengage team that guided us every step of the way. Special thanks to Mary Clyne, Product Manager, Larry Main, Managing Editor, and Michael Tubbert, Senior Content Project Manager.

It is our hope and dream that this work inspires the next generation to dream about flight and space exploration, that it motivates them to pursue participation in aerospace related fields, and that it is an engaging adventure

Review Panel

The publisher wishes to acknowledge the contribution of our dedicated review panel:

Brian Benton, Walton High School, Marietta, GA
Aaron Clark, North Carolina State University, Raleigh, NC
Bryce McLean, Coronado High School, Colorado Springs, CO
Dara Randerson, Oswego East High School, Oswego, IL
George Reluzco, Mohonasen High School, Schenectady, NY
Charles Spangler, Colonie High School, Colonie, NY
Ben Spenser, Jefferson High School, Lafayette, IN

We also wish to thank our special consultant for this series:

Aaron Clark, North Carolina State University, Raleigh, NC

And we especially thank Project Lead the Way's® curriculum director for engineering, Sam Cox, for reviewing chapters at the manuscript stage.

ABOUT THE AUTHORS

Ben Senson has been an educational innovator for 20 years, having developed the MMSD Remotely Controlled Observatory and its online training program, published activities for the National Project WET initiative, created and taught an aircraft construction experience course, and been a curriculum writer and master teacher for Project Lead the Way's Aerospace Engineering course.

Ben is also the lead mentor for the BadgerBOTS Robotics program, which operates Wisconsin's largest summer robotics camps, Wisconsin's largest FIRST LEGO League (FLL) program, the Badgerland Regional FLL Tournament, and a Chairman's Award winning FIRST Robotics high school competitive team. The Badger-BOTS most recent collaborative project involves the creation of a student designed and run factory to produce "sit-skis," an adaptive technology that allows athletes with disabilities to participate in cross-country skiing activities. Mr. Senson is the coordinator for the Global Academy, an accelerated career pathways magnet program for juniors and seniors offered through a consortium of eight school districts. His day job is teaching science at James Madison Memorial High School in Madison, Wisconsin.

Mr. Senson holds a bachelor's degree in secondary education from the University of Wisconsin - Madison specializing in the earth and space sciences and physics, a master's degree in astronomy and physics education from Ball State University, and is pursuing a second masters degree at this time.

Jasen Ritter teaches science and engineering at Vandegrift High School in Austin, Texas. He has been in education for more than ten years. Jasen is a Master Teacher for the Project Lead the Way Aerospace Engineering course and was the program developer for the Project Lead the Way/NASA Lunar STEM Enhancement Robotics curriculum.

In addition to teaching physics, Mr. Ritter is the lead mentor for the VHS Engineering and Robotic Programs and team mentor for the Computer Science Program. At Arlington High School in Arlington, Texas, he was the lead mentor for the Team America Rocketry Challenge, First Tech Challenge, and Engineering/Robotics teams. He was nominated as an AWARE Foundation Finalist for technology in the classroom and served as an advisor for the Fort Worth Chamber of Commerce Aerospace Consortium to help bring local industry into the classroom.

Prior to teaching, Mr. Ritter worked as IT Director for a national sports marketing agency. He holds a bachelor's degree in psychology and computer science from Florida State University, and a master's degree in education from Florida State University.

Aerospace Engineering:

From the Ground Up

An Introduction and Overview of Aerospace Engineering

GPS DELUXE

START LOCATION	DISTANCE	END LOCATION

Menu

Before You Begin

Think about these questions as you consider the concepts in this chapter:

1. What is an aerospace engineer?

2. What subjects are important for becoming an aerospace engineer?

3. What importance does teamwork play in engineering design?

4. How will aerospace engineering shape our future?

Aerospace engineering is one of the broadest of the engineering disciplines. An aerospace engineer must have a solid background in physics, chemistry, mechanical design, fluid dynamics, thermodynamics, and mathematics. Aerospace engineers play an essential role in our quest to explore the world and to bring distant sites closer to home. The various disciplines of aerospace engineering and other fields of engineering interact to create the engineering team, which ultimately brings the complete product together.

Whenever we open a new frontier to exploration, we rely on the skills of engineers to develop a craft that can carry us to our new destination. Engineers design the equipment that will protect us from danger on the journey and help us gather information about our new environment. Engineering innovation also helps us adapt to new environments so that we can explore them more fully with each returning trip.

The government, the military, and civilian companies develop aircraft and spacecraft to provide economic, scientific, and recreational opportunities that are unique to the aerospace environment.

This chapter just scratches the surface as we introduce the diversity of skills an aerospace engineer must develop and the many related subjects that make aerospace engineering such an interesting career opportunity.

THE AEROSPACE ENGINEER

The primary focus of the aerospace engineer is the design, construction, testing, and evaluation of craft that move through the atmosphere or outer space. This broad focus includes vehicles as simple as a sled, bicycle, or automobile and as complex as a high performance fighter aircraft such as the F-22 Raptor, or spacecraft such as the Space Shuttle or SpaceShipOne.

Most aerospace engineers specialize in one of the four major disciplines that must be understood to design a successful vehicle or enable it to safely operate: aerodynamics, avionics, materials science, and propulsion. Other engineers from outside disciplines, such as mechanical engineering, chemical engineering, computer science and even industrial engineering, will come together to make up the engineering team that assists in aircraft design.

Aerodynamics

Aerodynamics:

the study of the effect of air flowing over and interacting with objects and the production of forces such as lift and drag.

Aerodynamics is the study of how motion through a fluid affects both the vehicle and the fluid. When we think about an airplane, it's important to know that the craft must generate enough lift to pick itself up against gravity's pull. But it's just as important that we can predict how much of our power and energy will be wasted when forces such as friction and drag slow the aircraft's passage (Figure 1-1). Today's engineers spend a lot of time refining their designs to ensure that the resulting product is as efficient as is possible.

Figure 1-1 MIT/NASA Langley Magnetic Suspension/Balance System.

Source: NASA Headquarters—NASA Langley Research Center

Design Process and Product Development

The design process is a logical method for solving problems. You may already be familiar with the design process from your other engineering classes, and it can take many forms or steps depending on the application involved. For example, a typical aerospace design process and product development may be something like the following:

1. Identify customer needs.
2. Perform design planning:
   ▶ Concept design
   ▶ Preliminary design
   ▶ Detail design
3. Create a prototype.
4. Test the prototype.
5. Evaluate to ensure satisfactory completion of customer needs (repeat process).

As an example, identify the needs and compare the designs of various aircraft. Try a civilian Cessna 182 and the Lockheed Martin F22 Raptor. Why are wings high on top of a Cessna 182? What is the skill level of the user of the aircraft? What is the primary role of the aircraft? What other processes have you used in your problem solving that are similar to the design process?

Avionics

Avionics is the science of electronics and electrical systems. Aerospace engineers focused in this area spend most of their time designing the systems that enable people to communicate with each other, gather information about how the craft is performing, and display information to the pilot and other controllers (Figure 1-2).

Avionics:

the study of the electrical and mechanical instrumentation that indicates the status, condition, attitude, and future condition of a flying vehicle. Avionics includes controls for communications, navigation, engines, and aircraft movement.

Figure 1-2 *German test pilot Quirin Kim in a T-38 with a GEC-Marconi Avionics HMD helmet.*

Source: NASA Headquarters—NASA Dryden Flight Research Center

Figure 1-3 *X-29 cockpit.*

They also design human interface devices for command and control of the craft as well as for navigation and traffic avoidance (Figure 1-3). Avionics combines elements of science, technology, engineering, and mathematics with the human elements of psychology and physiology.

Materials Science

An understanding of materials science allows us to design a craft that is strong yet light, rigid but flexible, and durable but repairable, as well as safe for use in the construction, maintenance, repair, and disposal stages of the craft's life cycle (Figure 1-4). This field has played an ever-increasing role in the design of successful

Figure 1-4 *Vapor Crystal Growth System Furnace with Crystal.*

craft as we have evolved from the use of wood, to metal, and finally to composite materials for the construction of both aircraft and spacecraft.

Propulsion

Propulsion is the science of the engine. Whether we are relying on the downward pull of gravity, an internal combustion piston engine, a jet, or a rocket, propulsion represents the ability to take off, climb, and sustain flight over a long range (Figure 1-5). Throughout the history of aviation, improvements in engines have paralleled the expansion of our aircraft capabilities.

Propulsion:
the system or engine used to transform energy into a mechanical force. Propulsion forces are used to overcome drag and to stabilize and maneuver the craft.

Figure 1-5 **The Wright Vertical 4 engine powered the US Navy's first aircraft in 1912.**

Source: Ben Senson

Combined, these disciplines represent a broad understanding of all of the focus areas within engineering. Successful aerospace engineers possess an excellent understanding of the variety of challenges presented by a project, and they are skilled at combining a wide range of solutions to achieve a goal.

WHAT AN AEROSPACE ENGINEER UNDERSTANDS

This book introduces you to the broad range of subjects that might be a part of your future if you choose to become an aerospace engineer. We start with the history of aviation and the human drive to explore space. We explore the forces that air generates as it interacts with our craft to create lift and sustain flight. We also show you how engineers can use these forces to position and maneuver the craft so that we can truly fly to any destination we desire. We take a close look at propulsion: How do we get energy out of a fuel to create propulsion forces, and how do we channel that energy to propel the craft through air and space?

When we consider materials science, we look at how materials are used to construct a shaped piece of the aircraft or spacecraft, and we consider how individual parts can be assembled into a structure that serves a purpose within the large vehicle. We look at wood, metal, and composite materials and the role that they have played in designs of past and present craft.

Purpose Drives Design

As in all engineering disciplines, improving a design in one area frequently necessitates giving up performance in some other area of the design. Should a component be stronger or lighter? Can it be both? If we install a larger engine, we can have more power to go faster, but the aircraft will be heavier and larger, and will consume more fuel (Figure 1-6). The job of the professional engineer is to take into account the purpose for which the product is being designed and to ensure that the final design meets or exceeds the requirements without sacrificing performance, safety, or reliability (Figure 1-7).

Figure 1-6 SR-71 over snow-capped mountains.

Source: NASA Headquarters—Greatest Images of NASA (NASA-HQ-GRIN)

Figure 1-7 SR-71 Research engineer Marta Bohn-Meyer.

Source: NASA Headquarters—NASA Dryden Flight Research Center (NASA-DFRC)

Rocket Propulsion: Solid or Liquid?

The tradeoffs between two designs can be dramatic. For example, we can use a rocket motor to create the thrust to maneuver a spacecraft. Using a solid rocket has the advantage of being ready to fire at a moment's notice because its fuel is stable and easy to store. Liquid-fueled rockets have the advantage of being able to be turned on and off repeatedly, but their fuel is more challenging to store. So which one should be used? Aerospace engineers know that both are useful and rely on their experiences with them in the past to help choose which is the best solution for the current mission design.

The laws of physics remain the same for military and civilian craft, but the operational requirements for military aerial objects and vehicles can lead engineers to make design decisions that result in starkly different products. Missiles, guided precision bombs, and shells are all built for one-time use. Many military designs must be stored and ready for use on a moment's notice. The design might have to

Careers in Aerospace Engineering

READY FOR LAUNCH

NASA's space program is a risky operation, but it's a little less so thanks to Harmony Myers. Myers is a safety engineering section chief, supervising 12 other engineers who keep an eye out for bugs in the system. "We look for things that could cause harm to the Space Shuttle or the astronauts," she says. "We try to eliminate or minimize them so that it's a safer flight program."

On the Job

NASA launches the Space Shuttle four or five times a year, and Myers and her team are right there in the control room. She talks to NASA engineers about potential problems with the shuttle's hardware. If the launch doesn't meet requirements, she recommends that it be postponed.

Pilot James Barrilleaux enters the cockpit of a NASA ER-2 aircraft. ER-2s are civilian versions of the military U-2S reconnaissance aircraft. As part of NASA's Airborne Science program, ER-2s can carry scientific payloads of up to 2,600 pounds to altitudes of about 70,000 feet to investigate such matters as earth resources, celestial phenomena, atmospheric chemistry and dynamics, and oceanic processes.

Source: NASA/Ames Research Center

Myers is also involved in the prelaunch stage. She examines the shuttle's booster segments, looking for parts that might fail. She also checks out the system that monitors leaks in the shuttle's hydrogen and oxygen lines.

"The job is very challenging," Myers says, "and it changes day to day. Knowing you're part of getting the shuttle off the ground is inspiring. Every day, you're just grateful you can contribute to something wonderful like that."

Inspirations

Myers always loved math, and she was always curious about the way machines worked.

"I liked to take things apart and put them back together, like telephones," she says. "My parents probably hated me." In high school, Myers had a chance to shadow a professional for a day as part of a career program. "There was a list of professions, and I picked an engineer," she says.

She immediately knew that engineering was the career for her.

Education

Myers earned a Bachelor of Science degree in electrical engineering at the University of Central Florida and a master's degree in industrial engineering at the University of Miami.

"You learn a lot of math and science in college," Myers says, "but the main thing is to learn how to think analytically and logically through a problem."

As an undergraduate, Myers was inspired by the chance to work with Walt Disney World Ride and Show Engineering on a school project. She and a team of students were asked to work with a cable box used in Disney hotel rooms.

"It was really neat," Myers says. "We were given the box, and it was up to us to create an interface. We wired it to the thermostat so you could change the temperature in the room from your TV. It was a chance to brainstorm and figure out how to solve a problem on your own. That helped prepare us for the future of solving problems."

Advice to Students

Myers suggests that students talk to professional engineers about their jobs, just as she did. She also recommends taking as much math as possible, as early as possible, including calculus.

"As an engineer, you get to come up with solutions on your own to help better the environment, or to make things faster and quicker," Myers says. "And you can see your results. It's a very rewarding career."

Source: NASA Dryden Flight Research Center

Figure 1-9 Compton Gamma-Ray Observatory.

Source: NASA Headquarters—NASA Marshall Space Flight Center (NASA-MSFC)

withstand extreme cold, heat, and moisture, and some must be launched from the ground, an aircraft, or from beneath the sea (Figure 1-8).

Aircraft designed for outer space pose similar engineering challenges. Few of our spacecraft designs ever return to the surface of the Earth (Figure 1-9). A design is fundamentally different when the product can't be recovered and has to operate in the heat, cold, and intense radiation of outer space, where every maneuver draws from your limited supply of fuel on board.

THE FUTURE OF AEROSPACE ENGINEERING

The diversity of jobs taken on by aerospace engineers is staggering. They design and manufacture everything from missiles to NASCAR vehicles (Figure 1-10), from stunt planes to giant airliners, and must overcome challenges posed by materials, structures, interacting systems, and the anatomy, physiology, and psychology of the humans who operate their products (Figure 1-11).

Figure 1-10 Aerospace engineers work on far more then just airplanes. This model car is being wind tunnel tested to determine its aerodynamic characteristics.

Source: Jason Ritter

Figure 1-11 The updated glass cockpit of the Space Shuttle Atlantis.

Source: NASA JSC—NASA Johnson Space Center

Figure 1-12 ERAST Program Proteus aircraft.

Source: NASA Headquarters—NASA Dryden Flight Research Center

Figure 1-13 The Altair autonomous research vehicle is a modified military Predator B aircraft.

Source: NASA Dryden Flight Research Center

As the science and technology of aerospace engineering evolve in the future, it is important to consider current trends in aviation as well as the fundamentals of what has come before (Figure 1-12). Aerospace engineers rely on the knowledge gained by analyzing the performance of existing vehicles and investigations of past accidents so that the current generation of designs produces the most efficient and safe vehicles possible. However, the design constraints for modern aerospace vehicles are rapidly changing. Therefore, you will also find an introduction to autonomous vehicles in this book because they play an ever-increasing role in military (Figure 1-13), scientific (Figure 1-14), and civilian commercial aviation.

So if you're ready, fasten your seatbelt, and enjoy the ride.

Figure 1-14 The Space Shuttle glide test vehicle never flew in space.

Source: Ben Senson

SUMMARY

- Aerospace engineers possess a broad range of engineering skills. Areas of focus include aerodynamics, avionics, materials science, and propulsion. Aerospace engineers work in teams with other engineers to achieve a successful design in a timely manner.

- Aerodynamics focuses on the interactions between the air and an object as they move past each other. The result is forces that can be managed to achieve a goal of stability or maneuverability and to improve efficiency.

- Avionics includes all of the electronics and computer programming required to gather and display information about how the vehicle is operating and to control everything from communications, navigation, engines, and aircraft maneuvering through intuitive and reliable systems.

- Materials science studies the size, shape, strength, safety, and failure characteristics of every component in a vehicle. By understanding materials during the design process, we can improve the safety, functionality, and reliability of the final vehicle design.

- Propulsion includes both the engine that transforms energy from one type to another as well as the ability to transfer it to the craft so that it can achieve flight, sustain flight for a long duration, and maneuver to any chosen location.

- Aerospace engineers are applied scientists that use their understanding of the basic principles of physics, chemistry, and biology to enhance our ability to explore beyond the normal range of human capabilities and to routinely operate in the air and space environments without significant risk.

- The aerospace engineer can play a major role in our ability to be successful in warfare. Craft can be designed to be faster, be more maneuverable, carry heavier payloads, or be invisible to defensive systems. Aerospace engineers understand that achieving one design goal frequently requires sacrificing performance in some other aspect of the aircraft design.

BRING IT HOME

1. How might the skills of an aerospace engineer be useful outside of an aviation career?
2. What types of power plants are used on the vehicles that you use in your everyday life? Can you describe how they operate?
3. What interests you most about becoming a part of the aviation community?
4. Explore one of the major focus areas of aerospace engineering on the Internet. What types of applications did you find for the skills of an engineer specializing in that focus area?
5. Find at least one way that aerospace engineering has had a direct impact on your life. Explain the connection.

EXTRA MILE

1. Talk with a guidance counselor about postsecondary education and other opportunities in aviation-related fields.
2. Contact the Experimental Aircraft Association (EAA) to find out when free Young Eagles flights will be available near your community. Share this information with friends and classmates.
3. Take an introductory flight lesson at your local flight school.

CHAPTER 2
The History of Human Flight

GPS DELUXE

Menu

START LOCATION   DISTANCE   END LOCATION

Before You Begin

Think about these questions as you consider the concepts in this chapter:

1 What did the first attempts at human flight look like and why did they fail or succeed?

2 What are the four essential characteristics for a heavier-than-air aircraft?

3 Who are the Wright brothers and what did they add to our understanding of flight?

4 What changes in materials and design occurred during the "golden era" of aviation from 1920 to 1940?

5 How do jet and rocket engines operate, and which is more useful?

6 Why was there a sound barrier, and if it was a barrier, how was it overcome?

7 How do the roles humans and autonomous vehicles play in space exploration compare to each other?

8 What advantages and disadvantages do unstaffed aerial vehicles (UAVs) have over conventional aircraft?

9 How has competition and cooperation between nations affected space exploration?

10 What is the importance of continuing our effort to explore the Universe, return to the Moon, or build long-term permanent colonies?

Achieving mastery over flight has been a constant struggle throughout human history. From our first recognition that birds and insects can willingly travel through the air, we have desired the same freedom for ourselves. To appreciate the significance of a particular innovation in our understanding of flight, we need to understand the knowledge upon which it was built and the impact the innovation had on its time. The field of aerospace evolved from barely a powered flight to humans on the Moon in less than 75 years. How impossible lunar travel must have seemed to the Wright Brother's given the effort of their achievement and the technology they had at the turn of the 1900s. The calculator you use in class has more computing power than the entire computer system on Apollo. Space Shuttles, UAVs and the modern technology they are based upon did not spontaneously appear. Only through years of necessity and invention was the progression of aerospace technology possible.

Wilbur and Orville Wright's first flight on December 17, 1903 lasted about 12 seconds and covered 120 feet. On May 14, 2010, NASA's Space Shuttle Atlantis crew embarked on STS 132. Their mission lasted 12 days and covered 4.8 million miles.

This chapter takes you through the broad strokes of the history of aviation from the birth of lighter-than-air ballooning and the realization that humans could be unbound from the surface of Earth to our ultimate travel in outer space. The story has taken many detours along the way, but this telling will stick to the main road. Our goal is not to be exhaustive in discussing every innovation and airplane but instead to highlight the most important developments and aircraft that shaped history and guided it toward the next step in traveling higher, faster, and farther.

FLOATING ON AIR: ACHIEVING LIGHTER-THAN-AIR FLIGHT

The birth of human flight was not achieved by flapping wings or gentle glides. After hundreds of years of injury and death by foolhardy tower jumpers, Joseph and Etienne Montgolfier tripped upon the simplest of all methods for becoming airborne; to float upon the air itself. What inspired this insight? And why had no one thought of the idea earlier? Actually, someone had.

Everyone knows that air can be pushy; just step outside on a windy day. For centuries, people have made use of wind power to run their mills, run their machinery, and push around their ships. What is amazing about balloons, dirigibles, and airships is that they can fly in no wind at all.

As early as 200 BC, the great Greek scientist Archimedes had clearly defined the requirement for producing buoyancy, or lifting force, on an object: *simply exist*. To exist is to push other things out of the way, to take up space. The lifting force on any object is equal to the weight of this displaced matter. The value of the principle was immediately obvious for designing ships and other devices that would float in the relatively dense fluid of water.

However elegant, the principle remained nearly useless to the aerostatic investigator. Air has such a low density that the buoyancy force on a solid object is destined to be far smaller than the force of gravity and the weight of the object itself. What was required for flight was a vehicle that was itself lighter than air.

In 1670, a monk named Father Francisco de Lana Terzi proposed the best possible solution. What could be lighter than nothing at all? Terzi's aerial ship would have every possible molecule drawn out from four hollow copper spheres of 14 ft diameter. The result would be a displacement of almost 5,800 cubic feet (cu ft) of air. Air at sea level and 70°F has a density of approximately 0.0745 lbs per cu ft. From the spheres alone, Terzi's aerial machine could weigh as much as 427 lbs, and it would be lighter than the air that it displaced. Unfortunately, creating a perfect vacuum (no particles at all) is impossible, and Terzi overlooked the fatal flaw in his design.

When you remove gas particles from a container you reduce both the weight of the contents and the number of particles that are bumping into the inside surface of the container. This means that the more particles you remove, the lower the pressure on the inside of the container. However, you have done nothing to the pressure on the outside of the container. The lighter the craft gets by removing air from within the spheres, the greater the net force squeezing the sphere. As shown in Figure 2-1, the copper spheres to contain Terzi's vacuum could be built strong enough to withstand

Figure 2-1 (a) Terzi's design for an aerial ship relied on creating a vacuum to reduce weight, but it provides no gas pressure to resist crushing by the atmosphere. (b) Montgolfier's hot air balloon relied on warming the air inside the balloon to reduce its weight and create the gas pressure required to balance out atmospheric pressure surrounding the balloon.

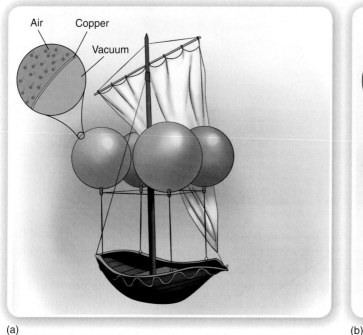

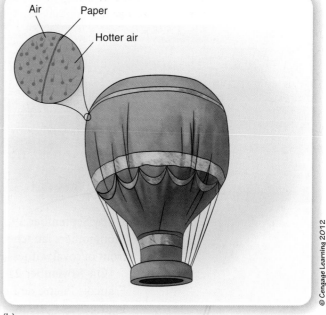

(a)

(b)

the crushing force of atmospheric pressure but not without their weight greatly exceeding the lifting force they generated. A lifting gas was needed that was less dense than air but capable of pushing back with air pressure equal to that of the outside air.

There are two ways to generate higher pressure:

▶ Increase the number of particles bumping into the inner surface.

▶ Increase the speed of the particles so that they bump into the surface more frequently and with greater force.

The second solution is the key to a Montgolfier balloon.

A Lot of Hot Air: The First Lighter-Than-Air Craft

In 1782, the Montgolfier brothers noticed that the ash from a fire rose upon the smoke and hot air rising from a fire. With a crude frame of wood and taffeta fabric, they built the world's first lighter-than-air craft designed to trap the heated air above a fire. The heated air was less dense but maintained its ability to push against the envelope's collapse. The balloon filled with hot air and floated to the ceiling of the room. The challenge was set; how to lift a human aloft with a hot air balloon.

The technological breakthrough was to create an airtight container that was large enough to produce the lifting force required and sealed well enough that the warmed air could not easily escape or cool. If 1,000 cu ft of hot air could produce 11.8 lbs of lift, then 10,000 cu ft would produce 118 lbs of lift. The Montgolfier family had exactly what was required because their family business was papermaking.

By June 4, 1783, the brothers had designed and built a 28,000 cu ft envelope out of linen sackcloth for strength and rag paper lining to contain the gas and insulate it against cooling. The cloth was held together by more than 1,800 buttons and reinforced by ropes that supported a smoldering brazier below the envelope. When launched, the balloon flew more than a mile and stayed aloft for almost 10 minutes.

Figure 2-2 First Flight of Passengers, September 19, 1783, in Annonay, France.

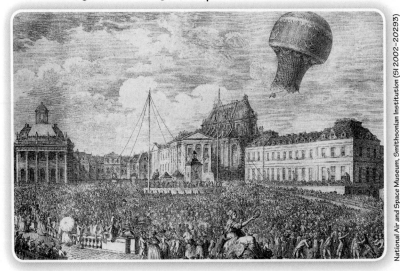

National Air and Space Museum, Smithsonian Institution (SI 2002-20293)

By September 19, the dimensions had grown to over 41 ft in diameter, and the Montgolfiers launched the first aerial passengers (a sheep, a duck, and a rooster) in front of royal witnesses King Louis XVI and Marie Antoinette.

On November 21, 1783, the Marquis d'Arlandes and the King's historian Jean François Pilâtre de Rozier became the first humans to fly untethered in public (see Figure 2-2). "The machine, say the public, rose with majesty . . ." wrote d'Arlandes. The flight lasted less than 25 minutes, and although the balloon endured holes burned through the envelope, several snapped supporting ropes, and a crash into the countryside, humans had achieved free flight!

Another Solution to the Problem: Hydrogen

Competition was afoot; Jacque Charles had developed an entirely different technology based on the same buoyancy concept of using a lighter-than-air lifting gas in an envelope. However, unlike the Montgolfier brothers, he used a gas that had a much lower density at normal temperatures; hydrogen. Within weeks of the Montgolfier's famous flight, Jacque Charles flew his first hydrogen balloon into the Paris countryside. Upon landing, the locals attacked the devil from the sky with pitchforks, but the proof of concept had already been made (see Figure 2-3).

Figure 2-3 Jacque Charles' hydrogen balloon is attacked by villagers.

National Air and Space Museum, Smithsonian Institution (SI 2002-20291)

Point of Interest

The Robert Brothers

Hydrogen had been discovered in 1766 as Phlogiston, or inflammable gas. By 1781, its lifting power had been demonstrated by creating soap bubbles filled with the gas. However, the gas had only been produced in small laboratory quantities. The technological breakthroughs that allowed Charles to use this low-density gas were made by a pair of instrument makers, the Robert brothers (Marie-Noël and Anne-Jean). Within a short period of time, they developed a technique for producing large quantities of hydrogen. They would fill oak barrels with iron filings and pour sulfuric acid over them. The escaping gas was piped into the balloon. By setting up a ring of five barrels around the balloon, they could recharge each barrel in succession and produce as much hydrogen as was necessary.

The challenge with hydrogen, once it is produced, is containing it. Hydrogen is the smallest atom on the periodic table. Thus, it easily escapes from almost any container. The Robert brothers solved this problem as well. They developed a technique to dissolve India rubber into turpentine. When painted onto a taffeta fabric, the coating rubberized the fabric and made it nearly gas tight. The lifting gas worked so well that Charles only needed balloons of 26 ft diameter to carry two men aloft. About 100 years later, at the beginning of the Civil War, Thaddeus S. C. Lowe would use this same technique to seal his hydrogen balloons used as observation platforms (see Figure 2-4)

Figure 2-4 Thaddeus Lowe's Intrepid is filled with Hydrogen during the Civil War, 1862.

National Air and Space Museum, Smithsonian Institution (SI 2003-25383)

Flying in an early balloon was a harrowing experience. Jacque Charles himself only flew once and then swore off the practice. The challenge was that the craft drifted with the winds, and it was difficult to control your altitude. To rise into the air, you needed to be able to either heat the lifting gas—dangerous enough for an envelope made of fabric, paper, and rope but suicidal for a balloon filled with hydrogen—or reduce the weight of the balloon by dropping ballast weight from the balloon. To descend required cooling the gas by waiting for nature to run its course or by venting the precious lifting gas from the envelope. Many balloon flights ended the tranquility of the silent flight with a less than subtle crash landing.

By the middle of the 1800s, balloonist Charles Green had invented the simple dragline for controlling altitude. The concept is incredibly simple. A long heavy rope is left hanging from the balloon. When the balloon descends low enough, the rope drags on the ground, which means that the balloon is no longer supporting the rope and is lighter so its descent rate decreases. If the balloon rises higher it lifts more and more of the rope off the surface, becomes heavier, and thus slows down its ascent. Green used a simple, infallible technology to solve the problem of altitude control as long as the rope wasn't accidentally dropped overboard!

Lighter-than-air flight was becoming safer but certainly no less practical. What was necessary was a means to pilot the vehicle against the wind to a desired location at will. Throughout the late 1800s, many designers began to suspend a rigid frame beneath the gas envelopes to mount a motor onto the balloon. Thus, was born the dirigible, or airship. Transforming the airships from underpowered, overweight contraptions in 1853 into sleek maneuverable craft by the late 1890s was in the hands of Brazilian-born inventor Santos-Dumont.

Choosing Your Destination: The Birth of the Airship

Count Ferdinand Adolf August Heinrich von Zeppelin cast the image of the classic airship in the early years of the 20th century. In 30 short years, the airship was transformed from a curiosity into the capable craft of the LZ-127 *Graf Zeppelin*, which, in 1928, circumnavigated the world in 29 days. By August of 1929, transatlantic flights were occurring regularly. The great Zeppelins were capable of crossing the vast ocean in comfort, safety, and in much less time than was required by steamer ship.

Figure 2-5 *Hindenburg* in flight over the Hudson River and downtown Manhattan, 1936.

Source: Lambert/Archived Photos/Getty Images

Zeppelin took the simple nonrigid envelope that held its aerodynamic shape with gas pressure alone and designed semirigid and rigid framed airships of ever larger size. By May of 1936, the pride of the rising Nazi party, LZ-129 *Hindenburg* was ready for its maiden flight. Over 800 feet in length, displacing air with 5.5 million cu ft of hydrogen, and capable of flight at more than 80 mph, the *Hindenburg* was the greatest airship ever built (see Figure 2-5).

Helium had been discovered in 1885, so why was hydrogen still used? The world's largest suppliers of helium were the United States and the Soviet Union who both have large deposits and resources to allow for separation of helium from natural gas. Just prior to World War II, neither country was selling the valuable gas to a nation they viewed as dangerous. On the practical side, hydrogen is much less dense than helium and less expensive to produce. So Zeppelins remained hydrogen airships.

The Death of the Airship

A little more than one year later, on May 6, 1937, the *Hindenburg* sealed the fate of airships as the preferred method for flying. As the aircraft approached the Lakehurst, New Jersey mooring tower for landing, it burst into flames and crashed to the ground, killing 35 of its 97 passengers and 1 person on the ground. In less than a minute, the symbol of Nazi engineering was reduced to a scorched mass crumpled on the landing field. The spectacle had been recorded on film and broadcast by radio, and the technology was doomed (see Figure 2-6). Most observers blamed the flammable hydrogen for the accident.

Figure 2-6 *The Hindenburg's fateful day on May 6, 1937 in Lakehurst, New Jersey.*

Source: Lambert/Archived Photos/Getty Images

Point of Interest

Why Did the *Hindenburg* Burn?

The fabric of the *Hindenburg* was sealed with a material called dope. The dope contained an aluminum compound that can be extremely flammable. So which was it? Although not conclusive, the Discovery Channel's *Mythbusters* focused on the subject in Episode 70. Check out the show to see their results.

Throughout World War II to the present day, the lighter-than-air balloon has remained a recreational vehicle, and the airship has been relegated to long-duration reconnaissance and antisubmarine warfare or advertising and television platform duty. The modern image of the airship isn't glamorous international travel; it is that of the Goodyear, Fuji Film, and other nonrigid blimps.

BEFORE WE GOT IT WRIGHT: ACHIEVING HEAVIER-THAN-AIR FLIGHT

"Not within a thousand years will man ever fly," Wilbur Wright, two years before the first flight of the 1903 *Wright Flyer*, the first successful heavier-than-air aircraft.

The birth of a successful heavier-than-air aircraft requires only four elements to be brought together: a strong lightweight structure to carry loads, wings to create lift to overcome weight, propulsion to overcome drag, and mechanisms to control the orientation of the craft in space (see Figure 2-7). It sounds simple enough, after all humans have been witness to the power, ease, and gracefulness of insect and bird flight for our entire existence. Unfortunately for us, the air keeps its secrets well

Figure 2-7 The four forces on a bird in flight.

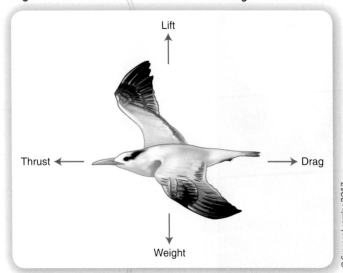

Lift

Thrust ← → Drag

Weight

© Cengage Learning 2012

hidden. This section follows the thousands of years it took for humans to invent the field of aeronautical engineering and to gain mastery over each of the basic elements of flight.

Leapers and Flappers: Failures, Foundations for Success

It is natural to assume that the uniformity of form and function seen in bird's wings is not by pure accident. The earliest "fliers" were careful observers of nature, constructing elaborate models scaled up to human proportions. They made use of feathers to ensure success. Although spectacular to the witnesses, the leap of faith from towers or cliff edges by the earliest aviators ended in broken bones or death (see Figure 2-8). They did not consider the limitations of the human engine for power output. The human body simply doesn't have sufficient muscle mass in the chest and arms to bear the load of flight. From John Damian's 1507 leap from Scotland's Stirling Castle to the Marquis de Bacquevilles Paris jump in 1742, the leapers were an aeronautical dead end.

Figure 2-8 Early flight scientists tried to simulate the flight of birds, often including feathers and wings in their designs.

© Cengage Learning 2012

But people are creative . . . experience has taught us that we can design mechanisms to amplify our strengths. In the 1500s, Leonardo da Vinci drew hundreds of sketches related to flight in his scientific (engineer's) notebooks. He was fascinated by constructions that today we would call parachutes, helicopters, and ornithopters (wing flapping aircraft) (see Figure 2-9). Although none of his designs would have been capable of flight, he is the first to consider the challenge of flight by looking at the physical processes behind the four forces of lift, weight, thrust, and drag.

Figure 2-9 *Leonardo da Vinci's helicopter. Models of his ornithopter mechanism are visible in the background.*

Source: William West/Staff/AFP Getty Images

Leonardo correctly proposed that pressure differences account for lift and drag forces, and that flight is relative to the air alone. The wind tunnel concept for testing flying machines rests on his ground-breaking assertion that, "As it is to move the object against the motionless air so it is to move the air against the motionless object" (*Codex Atlanticus*). Although da Vinci's discoveries could have been revolutionary, his work was insignificant in terms of bearing fruit through the discoveries of others due to his habits of keeping his notebooks secret and using mirror writing. What could have been major **innovations** in the history of aviation were not revealed until well after other investigators moved beyond these basics.

Math Informs the Inventor

By the late 1600s, Isaac Newton's monumental publication *The Principia* used mathematical theories to demonstrate that all bodies moving through a medium, such as air, produce drag forces proportional to the velocity squared.

Innovation:

a unique development or change in a product or service.

What's in a number? Whether a value is raised to the second power (squared) or third power (cubed) has a large impact on the resulting value. The classic example is the engineering relationship between the dimensions of an object and the weight of the object. Consider a really simple object, such as a cube, with all sides having a length of 1 ft. What happens when we double the size of the cube?

▶ The individual dimensions for length, width, height, and so on all double ($dimension^1$) by definition to become 2 ft long.

▶ The surface area of one face of the cube is length × width ($dimension^2$), so it increases from $1 \times 1 = 1$ to $2 \times 2 = 4$.

▶ The volume of the cube is length × width × height ($dimension^3$), so it increases from $1 \times 1 \times 1 = 1$ to $2 \times 2 \times 2 = 8$ cu ft.

▶ Weight varies with volume, so while the object is only twice as large, it weighs eight times as much.

Newton's work was intended to counter a prevailing theory by René Descartes that space was filled with matter and that there would be frictionless vortex-like motions created by the planets as they orbited the Sun. Newton's theoretical proof that all motion in a medium produces drag, and therefore objects must slow down over time unless there is a motive force, was important for this argument, but the fact that it was proportional to V^2 was essential for *designing* the performance of an airplane.

Newton, giant of 1600s science, got one thing monumentally wrong. In his formulation for calculating the lift and drag forces produced on a flat plate, he described them in terms of the sine and cosine of the angle of attack (AOA) of the plate. At low angles, the plate would produce very little lift. The obvious solution was to increase the AOA until sufficient lift resulted. According to Newton, however, the drag force would increase in force much more rapidly than the lift increased. Flight was impossible! Fortunately, this second conclusion was not only incorrect, but it was also largely ignored, and thus the tale of human flight goes on.

A Piece of Cardboard and a Box Fan

A simple experiment on lift and drag can be carried out with nothing more than a large piece of flat cardboard and a box fan. Hold the cardboard sheet horizontally in front of the box fan. Hold on tight as you slowly increase the AOA of your flat plate airfoil. Note the rapid increase in both lift and drag. Which force dominates at low angles of attack? How about at moderate and high AOA?

Figure 2-10 **Mariotte's dynamometer.**

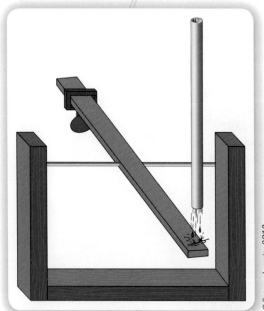

© Cengage Learning 2012

Edme Mariotte confirmed the V^2 relationship for drag in 1673 by use of his beam dynamometer and flat plates experiment (see Figure 2-10). Science was finally providing tools for the engineer investigating how to build a flying machine.

Inventing the Tools of the Trade

What was needed was a way to investigate other factors affecting lift and drag. But if you can't fly, how can you get data? In 1746, Benjamin Robins created two devices that served the purpose. His whirling arm and ballistic pendulum revealed two important aviation discoveries (see Figure 2-11).

Robins's whirling arm showed that it was more than just the volume of air being displaced by an object (a function of cross-sectional area alone) but that the shape of the object itself was critical. Testing triangular and flat-plate objects with identical cross-sectional area, Robins's data clearly showed that a triangular object produced much less drag when advancing into the air pointed end first. Also, rectangular plates with their longer sides as leading and trailing edges also produced less drag. This is the first description of the impact of aspect ratio on the efficiency of a wing.

The ballistic pendulum revealed that drag increased by V^3 when approaching the speed of sound. Robins accurately describes this transonic drag rise 200 years before any aircraft would challenge the sound barrier.

Figure 2-11 Benjamin Robin's whirling arm aerodynamics testing machine.

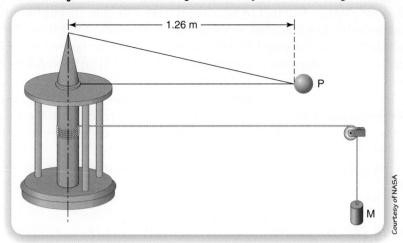

Courtesy of NASA

Robins's work was widely read by others in his field of ballistics, but it did not become part of the collective knowledge of aeronautical investigators. The final breakthroughs of the 1700s would come from exploring the efficiency of windmills and the flow of water in pipes.

Understanding the Force of Lift

In the 1700s, the available "engines" were wind, water, or muscle power. The great invention of the steam engine by Thomas Savery in 1698 would be nearly useless until James Watt improved upon the design in 1763. Steam engines were large, heavy, and expensive. Thus, there were windmills by the thousands throughout Europe providing power to grind grain, run pumps, and do other tasks. John Smeaton used flat plates oriented perpendicular to air flow to measure the force produced on them at a known velocity. To scale his results for an answer in pounds, he needed to insert a coefficient into the equation. Smeaton's coefficient allowed velocities in miles per hour and areas in sq ft to be used to predict the drag force that the flat plate would experience. The power of Smeaton's coefficient was that it then allowed all other investigators to calculate in advance and predict the performance of their design before it was actually constructed! Smeaton fixed the value of this coefficient at 0.005.

Daniel Bernoulli's name is associated with one of the most commonly stated theories of lift production. In the 1700s, he was investigating the relationship between the pressure and the velocity of water flow in pipes such as those rapidly spreading municipal plumbing throughout major cities. **Bernoulli's Principle** simply states that as the velocity of a fluid is increased, its ability to push against a parallel surface will decrease. It fell to his gifted colleague Leonard Euler to apply the new mathematical tool of calculus to derive the equation that we associate with Bernoulli. The equation has a simplified form that can be solved with simple algebra:

$$P + \frac{1}{2} \rho v^2 + \rho gh = constant$$

As the 19th century began, the young science of fluid dynamics was ripe for major innovations. Researchers Claude-Louis Navier and George Gabriel Stokes reduced the analysis of fluid flow to a series of equations that were nonlinear, coupled, partial differential equations. Their equations were nearly impossible to solve, but they were an indication that the subject could be understood in detail. Hermann von Helmholtz describes flow as having a vertical, or circulating, pattern. And Osbourne Reynolds, a name few nonprofessionals would recognize, described the changes required to force the flow of a fluid to change from that of low drag-inducing laminar flow to much higher drag turbulent flow. Reynolds also delivered one of the major breakthroughs of the entire history of aerospace engineering by describing the impact of the size of an airfoil on the resulting forces it can produce in the real world. The smaller a test model is, the faster the speed or the higher the density of the moving fluid at which it has to be tested to keep the same Reynolds number. To be of any value at all, we have to match the Reynolds number of a test object with that of the final aircraft for the test results to be accurate predictions of performance.

Preceding almost all of these developments, someone invents the modern airplane.

Defining the Airplane

In 1809, Sir George Cayley drew the world's first diagram of a fixed-wing aircraft with two perpendicular tail surfaces bolted to a fuselage. His drawing indicated an adjustable position for the empennage, thus allowing control of the forces on the tail and of the entire aircraft's orientation in space. In one moment, Cayley laid waste to the flapping arms of the past and laid out what is nearly a modern aircraft. Cayley had this diagram, as well as the first representation of lift and drag as components of one resultant force, engraved into opposite sides of a medallion. Sir George built the first whirling arm dedicated specifically to the study of flight, redefined Smeaton's coefficient to 0.0037, revealed that dihedral prevented the aircraft from rolling, and reinforced the evidence that higher aspect ratio wings produce much more lift compared to drag (L/D ratio). He even suggested that multiple layered wings, or decks, of high aspect ratio could be stacked to improve the efficiency of the airplane without requiring impractically long, weak wings. He is the father of the Tridecker, and of all biplanes and triplanes!

The close of the 19th century brought aviation into the spotlight through the work of several individuals. John Stringfellow demonstrated a steam-powered triplane that would "fly" on a wire across the Crystal Palace of London at the first ever Aeronautical Society of Great Britain exhibition. In the late 1800s, both Felix du Temple and Alexander Mozhaiski demonstrated aircraft capable of a powered "hop" into the air, which although brief and uncontrollable, demonstrated that a heavier-than-air craft was actually possible. Francis Wenham transformed da Vinci's theory into the first practical wind tunnel, which inspired Horatio Phillips to discover that a cambered airfoil, with curve to its upper and lower surfaces, produces much more lift than a simple flat plate. Phillips kept meticulous engineer's notebooks and published his findings in the 1885 journal *Engineering*, protecting his right of discovery.

By 1884, Alphonse Penaud had designed a two-seat monoplane using cambered airfoils, a wing dihedral for stability, elevators for maneuvering the nose up and down, a fixed vertical stabilizer and movable rudder, an enclosed cockpit, a single stick flight control, instrumentation, retractable landing gear, and a tail skid. Every aspect sounds modern. So what was missing? Why don't we credit du Temple, Mozhaiski, or Penaud with the invention of the airplane?

To be considered a powered and sustained flight, an aircraft had to fly along a horizontal or rising path, without losing airspeed, to a point clearly beyond where the momentum of the launch alone could carry it, while being satisfactorily stable in flight (*The World's First Aeroplane Flights*, 1965 by C.vH. Gibbs-Smith). The Wright brothers accomplished all of this, and more, in four short years.

Cambered airfoil:
an airfoil shape that has a more pronounced curve on one half than the other; non-symmetrical halves. The result is usually a pressure difference between the two sides.

THE RACE FOR FLIGHT, GETTING IT WRIGHT

The beginning of the 20th century was the time for the invention of the successful airplane. Every element of the craft existed in some rudimentary form. Someone just needed to put together the entire package of a strong, lightweight structure; lift- and thrust-producing mechanisms; and a means to stabilize and control the motion of the craft as it flew through the air.

Before the Wright brothers, there were two investigators, Otto Lilienthal and Samuel Langley, who nearly beat them to the discovery. One would fail to death itself and the other to structural failure at the critical moment of flight.

The Scientist Engineers

Several key scientists and engineers laid the groundwork for the first successful powered flight. Some of these individuals inspired or corresponded with the Wright brothers while others were in direct competition.

Otto Lilienthal and Sustained Flight

In 1889, Otto Lilienthal published *the* book that was to be read by every aviation inventor, *Bird Flight as the Basis of Aviation*. Lilienthal clearly demonstrated that airfoil shapes with curvature (camber) outperformed flat plate shapes. Compared to Horatio Phillips's miniscule amount of data, Lilienthal collected reams of data proving this point.

In addition, Lilienthal took a unique approach to "crunching the numbers" when he analyzed his data. Anyone who has taken algebra knows what happens when the same value appears in the top and bottom of a fraction—it cancels out. Lilienthal collected his data at various angles of attack and at 90° to the airflow, and then divided the two. These force coefficients were simple ratios with Smeaton's flawed value and velocity canceled out! They were pure, relative measures of the efficiency of an airfoil for producing lift and drag. These are the beginnings of our modern lift and drag coefficients.

But more than being just a researcher, Otto Lilienthal was a flier. From the local park to a specifically constructed test hill, Lilienthal hurled himself into the air thousands of times. During the 1890s, his weight shift gliders grew in size and aspect ratio. By 1895, Otto had overcome the structural weakness of a long, thin, high aspect ratio wing with the same solution as everyone else: the biplane configuration. His gliders soared over distances of hundreds of meters while maneuvering at the pilot's discretion. Otto Lilienthal's designs were ready for the next step—to add an engine. However, before he could test this, he perished during a test flight crash due to the stall. A gust of wind lifted the nose of his glider high into the air, forcing it beyond the critical angle where it produced very little lift. Otto plummeted more than 50 feet to the ground below. On his deathbed, Lilienthal declared, "Sacrifices must be made." Aviation lost a giant to the limitations of a weight shift aircraft.

Octave Chanute and Bracing

Octave Chanute, a Chicago engineer, wrote, "It will be seen that the mechanical difficulties are very great, but it will be discerned also that none of them can now be said to be insuperable, and that material progress has recently been achieved toward their solution" (*Progress in Flying Machines*, published in 1894). Chanute, a railroad engineer, added his own innovation to the mix by demonstrating the lightweight and rigidity of the Pratt truss cell structure for bracing multiwing aircraft (see Figure 2-12).

Figure 2-12 Truss design of the Wright Flyer.

Drawing by Louis P. Christman, National Air and Space Museum, Smithsonian Institution (SI 2000-4488).

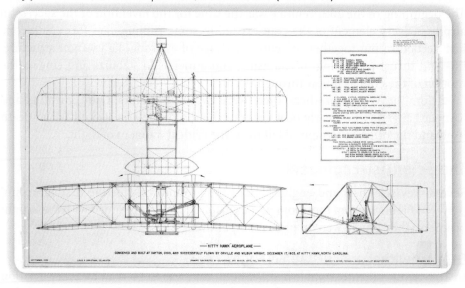

Samuel Langley and a Handful of Mortar

Samuel Langley was the third director of the Smithsonian Institution. He had made his mark as a solar astronomer but had been carrying out aeronautical investigations for several years by the time of Lilienthal's unfortunate demise. Langley refined Smeaton's coefficient to a value of 0.003 (today we know it is 0.0029) and precisely determined the impact of aspect ratio on lift versus drag. He also made what seemed to be a ridiculous discovery . . . it takes *less* power to fly faster! What Langley couldn't know is that he was testing his airfoils on the backside of the power curve. At extremely slow speeds, it takes a tremendous AOA to produce sufficient lift; therefore, the airfoil produces a lot of drag that demands a more powerful engine to overcome.

Langley built numerous small-scale models to test his theories. By the time he built Aerodrome #5, the models were capable of sustained flight. In May of 1896, the 13 ft wingspan, 26 lb, 1 HP model would be flown before witnesses. The craft was launched, as were all of Langley's airplanes, from a houseboat on the Potomac. After it initially settled lower toward the water, it began a stable climb until its engine literally ran out of steam. Aerodrome #5 achieved every goal except for maneuverability and the support of a pilot on board. Langley raced to take these next steps.

Only months before the Wright brothers achieved sustained, powered, manned, controllable flight, Langley's aircraft was hurled off the houseboat with Charles Manly, his assistant, on board. The craft met spectacular failure in front of the media. Both attempts, in October and December of 1903, sent the aircraft plummeting into the Potomac within 20 ft of the end of the launch rail. A reporter stated that it left the rail, "like a handful of mortar." Langley's marginal aerodynamics, longitudinal and directional control, and structural strength doomed the craft to failure. His major success was in the development of what was very likely the most powerful engine per pound of weight on the planet. The engine produced over 52 HP while weighing just over 200 lbs. Langley had predicted a need for only 12 HP. This throttle-less engine would likely have been too much for the airplane and would have accelerated it to such high speeds that aerodynamic forces would have ripped it apart. Eleven years later, Glenn Curtis would try to invalidate the Wright brothers' patents by demonstrating that Langley's design was airworthy, but only after 93 major modifications could he get the great aerodrome to fly at all.

The Wright Brothers

The Wright brothers are the inventors of the airplane. Not only were they the first to achieve success at the task at hand, but they also held the most complete understanding of why the airplane performed as it did. They were the first to design an airplane that performed exactly as it had been designed to fly. But what did Orville and Wilbur do that other researchers had not? What put them onto the path to success?

Kite Flying and Control

In May of 1899, Wilbur Wright sent a letter to the Smithsonian Institution to, "avail myself of all that is already known and then if possible add my mite to help on the future worker who will attain final success." In response, they were sent Chanute's *Progress in Flying Machines,* Langley's *Experiments in Aerodynamics,* the *Aeronautical Annuals* from 1895 to 1897, and four pamphlets by Langley, Lilienthal, and others. Through their studies, they realized that the pilot's weight would be an ever-smaller influence on larger airplanes. By the end of the year, the brothers constructed a biplane kite that relied on aerodynamic forces to achieve control rather than the shifting of weight.

Throughout the 1800s, stability in flight had been achieved in pitch and in direction through Cayley's simple design of two perpendicular surfaces at the tail of the aircraft and in roll by the use of dihedral in the wings. By moving the tail surfaces, the aircraft would be controlled in pitch and direction, but no one had any idea how to control roll without weight shifting. With their very first aircraft prototype, the Wright brothers cracked this problem with their innovation of the wing warping technique. If each wing is set to the wind with a different AOA, then each produces a different amount of lift, and one wing will lift while the other descends (see Figure 2-13).

Figure 2-13 *Octave Chanute's truss design adapted for the Wright Brother's wing warping control system.*

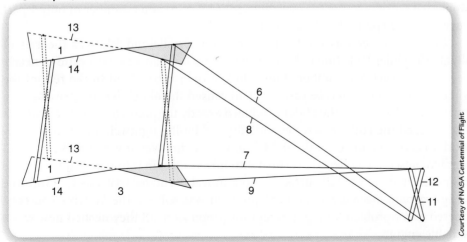

A Full-Scale Aircraft: Testing Lilienthal's Tables

In 1900, the brothers set out to design a full-size glider capable of carrying them into the air. They intended to gain flight experience with hundreds of hours of flight time. Instead, they revealed the limitations of the knowledge of the day. After searching the country for a suitable testing site, the National Weather Service suggested Kitty Hawk, North Carolina, for its sustained winds of 18+ mph. Using Lilienthal's tables of force coefficients, as published in Chanute's book *Sailing Flight*, the Wrights predicted a need for 200 sq ft of wing surface at a 3° AOA. With this configuration, they thought an 18 mph wind could sustain both plane and pilot in flight.

Unable to find suitable materials for building the spars for the glider, the brothers had to settle for an area of only 165 sq ft. The new size required a 21 mph wind. But the glider wouldn't fly. Only with 25 mph winds and an AOA of nearly 20° could the glider support itself and the pilot. Something was drastically wrong with the lift equations they had used to design the glider!

Return to the Basics: Gathering Evidence

In 1901, they built a much bigger glider using wings that matched Lilienthal's highly cambered wing design so that the predicted lift would be more accurate. The 240 lb glider had 290 sq ft of area. When flown, the glider was nearly uncontrollable. Full deflection of the horizontal control surface, now called an elevator, could barely keep the aircraft under control. The Wrights correctly realized that the location of the average lift was moving in the opposite direction of what everyone had thought would happen when the wing was put at different angles of attack. As the pitched up nose was brought down, the lift shifted forward at high angles of attack but then quickly moved to the rear of the wing at low angles of attack. This meant that the weight of the glider and pilot was farther and farther in front of the lifting force that was holding up the airplane. Just like a teeter-totter, this caused the heavy side (the nose) to swing downward. After numerous nose dives, the Wrights adjusted the airfoil shape to be less cambered, and the glider became controllable. But still, they achieved only a third of the predicted lift.

Although the Wright brothers likely misinterpreted Lilienthal's tables, they couldn't have known this. Rather than giving up, they did what any good engineer would do: They set up their own wind tunnel to gather their own lift coefficient data. After 1901, the Wright brothers had in their possession the most accurate data on airfoil performance in the world. They tested different cambers, aspect ratios, wing shape, and configurations such as biplanes and triplanes. By the end of 1901, they knew that the highest performance airfoil for the speeds they were attempting would have a span 6 times its chord length (AR = 6) and a camber of only 1 part curvature in 20 parts chord—a very flat airfoil!

The 1902 glider was the final test bed for Orville and Wilbur's theories of flight. The glider had almost the same wing area but had an aspect ratio of nearly seven and a very low camber. They added a twin vertical tail to the rear of the glider to overcome adverse yaw that had caused the old glider to steer against a turn. When test flown, the glider produced exactly the predicted lift. The brothers had cracked the code of aerodynamic lift and built a lightweight, strong framework to carry the forces of flight. The only issue discovered was that the new tail surface caused the glider to overturn into turns and spin into the ground. Orville proposed making the tail surfaces into a movable rudder that was tied into the wing-warping mechanism, and the problem was solved. The Wrights had conquered every problem of flight except for propulsion; all they needed now was a suitable engine.

Putting the Power in Powered Flight

Today, we take engines for granted, but in the early 1900s, they were anything but powerful, lightweight, or reliable. The Wrights wrote to every major manufacturer seeking a 9 HP motor that weighed less than 180 lbs and ran on the new fuel called gasoline. None could be found. With the help of their machinist, Charlie Taylor, they designed and built their own engine. An aluminum block with cast iron cylinders, convecting water for cooling, and a gravity-fed gas supply; this throttleless engine would produce 12 HP for 200 lbs of weight. After Langley's engine, this was the best engine in the world.

So, the question was how to transfer this power to the air. The Wrights assumed that there would be a body of knowledge on propeller theory; after all, humans had been making ship propellers for decades. What they found was a world of tinkers with very little theory. The Wrights realized that a propeller is nothing but a lot of individual airfoils moving through the air. The challenge was that as a propeller spins, the tips of the blade move much faster through the air due to rotation than does the root of each blade, yet both are moving forward through the air at the same speed. Each blade-element needed to be designed for its own best efficiency. When they were done designing and carving out their blades, they transformed over 65% of the power from the engine into power to overcome drag. This was leaps and bounds above that achieved by anyone else on the planet, but more importantly; they had a theory that allowed anyone to *design* a propeller for a task. The Wright's *Flyer* was ready to go.

Wright Success!

On December 14, 1903, Wilbur Wright could have become the first human to fly a powered, heavier-than-air, controllable aircraft (see Figure 2-14). Instead, the aircraft lurched upward and with an overcorrection slammed back down to Earth. Repairs took three days, and, by prior agreement, the brothers switched places. On December 17, with Orville at the controls, and a local resident on a camera, they started the engines, and the 1903 *Wright Flyer* made history. Four flights lasting no longer than 59 seconds changed the world forever.

Although the Wrights inherited the biplane with horizontal and vertical surfaces and wing dihedral for stability from Cayley, cambered wings from Phillips and Lilienthal, aspect ratio from Wenham and Langley, structural bracing from Chanute, gas engines from Langley, and propellers from Stringfellow, they are the first to design methods for control in all three directions, the first to understand that a propeller is a twisted wing, and the first to integrate everything into a system with just enough functionality to accomplish the task of sustained human flight.

Figure 2-14 *First flight of the Wright Flyer, December 17th, 1903.*

Courtesy of NASA

DESIGNING IN THE EARLY 1900s . . . HIGHER, FASTER, FARTHER!

In the opening decade of the 1900s, the airplane was manufactured in limited numbers. The first *Wright Flyers* were used for exhibitions and recreation while the Wright Model B of 1910 was intended for use by the military for pilot training and reconnaissance by the U.S. Army Signal Corps. The aircraft of the time retained the essential elements of the original *Flyer*: high aspect ratio biplane wings with a thin cambered shape. Pilots sat out in the open so that they could feel the wind in their face and fly by the seat of their pants.

War Breaks Out

Competition is never far behind the first successful proof of a new technology. Within a few short years of their first flight, the Wrights were openly challenged by the designs of Glenn Curtiss, Louis Blériot, and others. For designers everywhere, it all changed in 1914 with the outbreak of World War I. Suddenly the airplane had a life-and-death purpose. A pilot with an observer or a camera could suddenly reveal the plans of the enemy at the command of generals. Of course, it wasn't long before the military figured out that they could just as easily toss small grenades and bombs from the cockpit. It was a natural progression to carrying aloft firearms to fend off enemy aircraft. Before long, engineers developed aircraft specifically for the role of aerial combat, and the fighter and bomber were born!

Specialization: Bombers and Fighters

To carry a large bomb load, the aircraft had to become much more efficient at producing large amounts of lift without excess drag. The bombers of World War I were the largest biplanes ever constructed. By 1917, the German Gotha was capable of delivering 1,000 lbs of bombs onto London in formations of many aircraft. The bombers were slow but well defended by their movable machine guns and by the covering fire of the other bombers in the group. By 1918, the Italian Caproni Ca.42 heavy bomber could deliver more than 3,000 lbs of bombs per mission. Bombers were playing a major role in the war.

The development of improved bombers required better fighters to intercept and shoot them down before they delivered their load over the target. Fighters of the day had to be fast enough to catch the bombers and to outfly enemy fighters while being maneuverable enough to dogfight. While ever-increasing engine power allowed for faster flight, the addition of multiple wings with higher wing loading created the best fighters of the day.

Fun Fact

Goggles and a Silk Scarf: Fashion Statement?

The classic image of the pilot with goggles and scarf is a result of the era. The early engines were very unreliable and often overheated or seized. One solution was to build the engines looser and have the entire engine block rotate with the propeller to improve cooling. This worked fairly well, but it meant that a lot of unburned fuel and lubricant was thrown out the exhaust ports of the engine. Because castor oil was a common lubricant at the time, pilots needed the silk scarf to wipe the oil off their goggles, to cover their mouths to reduce the amount of oil ingested, and to reduce chaffing when they twisted around to track enemy aircraft. The classic scarf was far more than a fashion statement.

Throughout World War I, pilots sought to enter combat only when they had a clear advantage. Most of the aces of the war earned their ranks by shooting down unsuspecting or poorly defended targets. The unreliable wooden airframes and engines made aircraft of the day just as likely as combat to kill a pilot. The Sopwith Camel is an aircraft known for its difficult handling. It was common for other aircraft of this time to have parts of wings or control surfaces tear off during aggressive maneuvers as a result of the building materials.

After World War I, designers finally had time to return to basic research. Since the time of the Wright brothers, engineers had made use of the wind tunnel to gather data on airfoil performance. However, most designers knew very little about why the aircraft performed the way they did. The aircraft of World War I all looked fairly similar as designers borrowed design elements from each other as each airplane demonstrated what it could do in actual flight. The time was ripe for building the theory behind the practice.

THE GOLDEN ERA OF AVIATION

The 20-years between World War I and World War II represent a period in which the most rapid development of every facet of aviation occurred simultaneously. Aeronautical engineering theory, speed and distance records, military aviation, and engine technology all improved by leaps and bounds.

The Roots of Innovation

In 1915, the U.S. government founded the National Advisory Committee for Aeronautics (NACA) by an act of congress. The purpose of the organization, later NASA, was to support the basic research that would support commercial, civil, and military aviation. Through their pioneering endeavors, NACA engineers designed engine cowls that greatly reduced the drag produced by the exposed engines on an airplane while allowing sufficient cooling air to pass over the engine, which made aircraft much faster. NACA engineers carried out a series of organized airfoil tests based on the characteristics of the airfoil to create a national information base on airfoil performance that is still referenced today. And it was NACA, eventually NASA, that designed, built, and operated the largest experimental wind tunnels in the United States.

By the advent of World War II, aeronautical engineers had uncovered that the thin wings of the prior age were inefficient. Airfoils that were much thicker, with rounded leading edges, produced far more lift and less drag than their earlier counterparts. The advantage of the thicker airfoils was exploited to allow for stronger internal bracing of the wing. This, in turn, eliminated the external guides, struts, and bracing wires, which allowed the monoplane to dominate every area of airplane design. Thus was born the golden age of aviation in which numerous records and milestones were set by ever more capable aircraft and pilots.

When World War I ended, a flood of surplus aircraft became available for purchase by the returning pilots. The age of the barnstormers and stunt pilots was in full swing as informal air shows traveled across the country. The barnstormers exposed the public to aviation and gave many youngsters their first taste of flying. Throughout the 1920s, the age of the barnstormer slowly gave way to the age of the racers and record breakers.

Setting the Mark: Breaking Every Record

By 1919, the largest spectacle that could be used to demonstrate the capabilities of the airplane was a flight across the Atlantic Ocean. The first to accomplish the task was one of three U.S. Navy flying boats (the other two were forced down soon after takeoff) in 15 days. Within 2 weeks, a nonstop flight was made in less than 17 hours. Others flew the rim of the continental United States or to Alaska. By 1921, Bessie Coleman had earned the first pilot's license issued to an African American (she had to travel to France to receive training). In 1922, the helicopter H-1 was born with a short hop into the air for just over a minute of flight time. Pilots began to cross the United States in less than a day, and, by 1923, mid-air refueling demonstrated that the only limits to long duration flight were pilot fatigue and mechanical failure. By the end of 1924, fliers had completed around-the-world flights—after 175 days of travel! In a stunning move, the U.S. Postal Service opened Airway Route 1 from New York to San Francisco to support regular airmail service that beat the fastest trains of the era by days.

Commercial Aviation Takes Off

With two acts of congress, laws created the market forces that would pay commercial air carriers not just for carrying the airmail but also for the capacity of cargo they could carry while delivering the mail. After 1930, the subsidies from the government made it profitable to fly bigger, faster aircraft with large cargo and passenger-carrying capabilities. With Charles Lindbergh's 1927 nonstop solo flight across the Atlantic and Amelia Earhart's and Howard Hughes' rise to fame, commercial aviation gave birth to the airline industry. When Pan-American Airways began to offer transcontinental flights, other airlines raced to acquire airplanes capable of carrying passengers farther, faster, and on less fuel. United Airlines placed orders with Boeing Corporation for its model 247 while TWA and American Airlines convinced the Douglas Aircraft Company to produce the DC-2 and DC-3.

Figure 2-15 DC-3, the airplane that revolutionized commercial aviation.

The golden era of commercial aviation was borne on the wings of the *DC-3* (see Figure 2-15). Sleek, all metal fuselages built out of the latest material, a mixture of aluminum and copper called duralumin, allowed the aircraft to become enclosed and more streamlined, and to mount ever more powerful engines. With the advent of the DC-3, it became common to cross the continent reliably in less than 16 hours with only three stops for fuel. Flying quickly became the preferred method for traveling great distances . . . if you could afford the price. By the end of the 1930s, the DC-3 was transporting over 90% of the commercial traffic in the world. The DC-3 remains flying today and has been reborn as a modern turboprop-powered aircraft recertified as the Basler Turbo-67 (Figure 2-16). The Basler strongly resembles the classic DC-3 shape yet has more powerful turboprop engines, longer wings, and stronger fuselage than the original DC-3 and can be modified for special duties; the aircraft in Figure 2-16 is fitted with snow skids for delivering supplies to the South Pole research station.

Figure 2-16 Basler Turbo 67, an upgraded and modified turboprop version of a DC-3 airframe.

Military Aviation in the Golden Age

The growth of military aviation during the golden age can be directly credited to the work of General William Mitchell. His tireless demonstrations of aviation's impact in a time of war included naval attacks from "carriers," around-the-world flights, and aerial refueling. After numerous failed attempts to form a strong military aviation program, Mitchell openly criticized the weakness of American defenses, especially at Pearl Harbor where an aerial assault could cripple the fleet. Court-martialed and busted in rank, Mitchell wouldn't survive to witness his prophetic insight become a reality. However, soon after his death, the Army Air Corps was formed, and military aviation began to develop in earnest.

Building a Better Engine

Just as important as the transitions from biplane to monoplane, the change from thin to thick wings and improved aerodynamic design was the development of the aircraft engine. At the start of World War I, the Gnome rotary engine was capable of producing only 80 HP (see Figure 2-17). Although the engine was well balanced, produced uniform air cooling by rotation, and needed no flywheel to make it run smoothly, it had the distinct disadvantage of a large gyroscopic effect and a rapid use of both fuel and lubricating oil. The engine could only operate for 12-14 hours before it required a complete overhaul. The history of aviation has rested firmly on the development of capable powerplants since the Wrights were forced to design their own engine for the *Wright Flyer*.

Figure 2-17 *The LeRhone 9C rotary engine delivered 80 HP—similar to the Gnome rotary engine.*

Courtesy of Mark Miller

The initial gains in rotary engine performance were minor as reduction gears were added to allow the propeller to turn at a slower speed than that of the engine, which allowed a significant improvement in the efficiency of the propeller's ability to transform shaft horsepower into thrust. Far greater improvements in engine performance required a return to the prior design concept of a stationary engine block.

The radial engine looks deceptively similar to a rotary; however, unlike a rotary engine, the radial's cylinders and block stay stationary while only the crankshaft and propeller spin. Soon after WWI, the engineers had figured out how to redesign the cylinders to overcome the overheating problem that had prevented the engine's use during the Great War. The competing design of an inline engine also finally succumbed to a reengineering into the famous "V" engine configuration still seen today. Both types of engines quickly exceeded the rotary engine's limit of about 150 HP with power outputs ranging up to 1,000 HP by the end of the era.

If you set an aircraft from the end of World War I next to an airplane from the beginning of World War II, you can see the tremendous advances that took place in every facet of aerodynamic design, materials, and powerplant design (see Figures 2-18 and 2-19). The aircraft at the end of the golden age of aviation produced lift with little drag, had engines that were powerful and tucked into NACA engineered low drag cowlings, and had cargo capacities and flight performance that allowed them to fly higher, faster, and farther.

Figure 2-18 *The Gotha Bomber of World War One.*

© Cengage Learning 2012

Figure 2-19 *The B-17 Flying Fortress.*

Courtesy of United States Air Force

WORLD WAR II: METAL, MOTORS, AND MUNITIONS

The major innovations that are associated with World War II were born in the final years of the golden era. The innovations included major improvements in airframes, propulsion, communications and navigation, foul weather operations, and manufacturing.

The Need for Performance

The B-17 Flying Fortress is one of the best-known bombers of all time. The B-17 was developed by Boeing to meet a U.S. Army Air Corps call for an aircraft capable of 10 hours of flight at 10,000 feet and a maximum speed greater than 200 mph. With its 4 powerful engines and streamlined metal construction, the B-17 exceeded expectations for performance and set the standard for durability under fire. The legendary airplane became a favorite of aircrews on all fronts. The improved performance and range of the bomber created the need for fighters capable of long-range escort to targets deep in central Europe.

More powerful engines increased airspeeds and bomb loads, more efficient engines extended the range an airplane could fly, and more reliable engines increased the impact of an aircraft in combat. Jet and rocket engines were coming of age as well. Jet engines would be used in every type of aircraft from fighter to reconnaissance and the V-1 Buzz Bomb. The best example of the type is the all-metal Messerschmitt Me-262 fighter/bomber/interceptor, which was capable of speeds greater than 500 mph. The writing was on the wall . . . incremental improvements in mature internal combustion engine technology had raised fighter aircraft speeds above the 400 mph mark, a speed that was easily exceeded by the still rudimentary jet engine. Developed by Wernher Von Braun and German scientists and engineers, the rocket engine would play its role as well by providing the power for the guided V-2 long-range bomb and the Me 163 Komet. The V-2 rained terror on the British in the closing year of the war while the Komet pushed interceptor speeds beyond the 600 mph mark. Although too late to change the outcome of the war, jet and rocket engine technology defined the future of aviation innovation.

Making It Safe to Fly

During World War II it became much safer to fly. Within the United States, the federal government developed more than 535 new airports, emergency landing fields, radio navigation aids, and course and routing beacons. The newly developed RADAR and two-way radio systems allowed pilots to receive guidance to their targets and to talk to other aircraft in their formation and to ground controllers. The ability to see enemy aircraft on radar allowed limited resources to be used to greatest effect; enabled night time and foul weather interception of incoming attacks; and with the development of airborne RADAR units, allowed one-on-one independent interception of enemy aircraft. Of equal importance in foul weather was the development of de-icing equipment. In civilian aviation, flights are frequently delayed, rerouted, or cancelled due to icing conditions. Build up of ice on an aircraft increases its weight and decreases the effectiveness of the wings and control surfaces, which creates a rapidly deteriorating situation. De-icing equipment allowed aircraft to survive flight in much worse weather than ever before.

Rosie the Riveter: Changing the Manufacturing Process

Although seldom talked about, perhaps the most influential innovation of World War II had much more to do with manufacturing than with aeronautical engineering or flight performance. During the first three decades of aircraft manufacturing, almost all airplanes were built from beginning to end by a limited number of highly skilled craftsmen. The incredible number of aircraft necessary to wage modern war (some estimate well over 300,000 aircraft were produced by the United States and Great Britain during WWII) required a much faster manufacturing process. Manufacturing embraced the assembly line and broke assembly into much smaller, specific, easy-to-master steps that could be carried out by less skilled labor. In addition to changing the manufacturing process, aviation and industry also embraced the other half of the potential workforce. World War II would not have been won without the efforts of many "Rosies" and other women in industry.

⑥ BREAKING THE SOUND BARRIER

The British gave rise to the mythological sound barrier. In 1935, aerodynamicist W. F. Hilton pointed at a plot of the drag produced as an airfoil accelerated to high velocity (see Figure 2-20). Hilton said, "See how the resistance of a wing shoots up like a barrier against higher speed as we approach the speed of sound" (John D. Anderson, Jr., *The Airplane, A History of its Technology*).

Figure 2-20 The transonic gap.

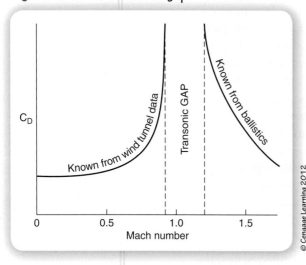

Ancient Roots

We have known since the time of Isaac Newton that sound has a finite speed. Anyone could observe the flash of a cannon's blast and then count until the report was heard. Today we know that at sea level, sound travels at a speed of nearly 1,117 feet per second (fps) or about 760 mph.

Although Benjamin Robins and his ballistic pendulum had allowed detection of the drastic increase in drag associated with approaching the speed of sound, this should not have led to the conclusion that supersonic speeds are impossible. We have demonstrated that the sound barrier can be broken as routinely observed in gunfire and in the crack of a whip.

Fun Fact

How Close Is the Storm?

When you see a lightning flash and start counting until the thunder is heard, for every second counted, the strike is about 1,117 feet away. So if you count 5 seconds between flash and boom, the strike is just a little more than a mile away.

Going Fast: Hints of a Bigger Barrier Problem

In the era of the wire and strut biplane, speed was an issue in only two places. In a dive, many aircraft could get going so fast that the forces of maneuvering the airplane could break pieces of the aircraft (not a good thing, but a result unrelated to the speed of sound). An object rotating has a much higher speed at the outer edge than at the hub, which is only rotating in place. The propellers of many airplanes could approach the speed of sound at their tips.

Throughout the golden era, scientists and engineers had witnessed a drastic reduction in propeller efficiency whenever the tips approached the speed of sound; however, this did not happen often. As engineers figured out new ways to reduce drag and increase the power output of engines and propellers, it soon became obvious that what was afflicting only the propeller would soon be impacting the performance of the entire airplane. In 1931, a Supermarine S.6B won the Schneider Cup race with a speed that exceeded 400 mph, or more than half the speed of sound. How much longer would it be before the airplane itself exceeded the speed of sound and went **supersonic**?

Although it can be quantified, the speed of sound varies with altitude, pressure, and a number of other factors. Thus, scientists and engineers find it useful to describe speeds relative to the speed of sound rather than as an actual speed. The **Mach number** (or Mach speed) is a ratio between the actual speed of the airplane and the speed of sound under its current flight conditions. Thus, a Mach number of 0.5 means the airplane is traveling at half the speed of sound, while a Mach 3 flight is at three times the speed of sound.

$$M = \frac{v_s}{u}$$

where

M is the Mach number.

v_s is the speed of the source (the object relative to the medium).

u is the speed of sound in the medium.

The Transonic Gap

By the 1930s, lift and drag were well understood for speeds up to Mach 0.8 and beyond Mach 1.2—at lower values due to aeronautical engineering research and flight performance of airplanes, and at the higher values due to ballistics and the study of bullets and cannon balls. The question was what would happen at nearly the speed of sound? Could an airplane gently accelerate through Mach 1?

What was known about transonic flight included the following:

▶ Most airfoils had a critical Mach number of 0.5 to 0.7 beyond which aerodynamics became transonic (see Figure 2-20).

▶ Beyond the critical Mach speed, lift rapidly plummeted by 20–80%.

▶ Drag skyrocketed by more than 400%, with no known maximum value.

▶ The center of pressure rapidly moved backward on the airplane, which made it nose heavy and prone to "tucking under" (a dangerous outcome as speeds increased in the resulting dive!).

▶ If the AOA or thickness of the wing increased, the critical speed at which the effects started was at a slower speed.

Designing a Supersonic Airplane

The next step required an aircraft dedicated to studying the nature of flight aerodynamics leading up to, and then exceeding, the speed of sound. During World War II, it became painfully clear that speed and performance dictated who would win in aerial combat. The Army Air Corps provided funding to design and build the aircraft required to complete the research.

The engineers wanted an aircraft that met five basic criteria:

1. Uses a small turbojet engine for propulsion
2. Takes off under its own power
3. Has a maximum performance speed of Mach 1
4. Has capability for a large payload of scientific instrumentation
5. Collects data throughout a slow process of incrementally approaching the speed of sound over an extended period of time

The Army Air Corps held the purse strings and demanded more; the captured documentation on German rocket engine research made it clear that rocket engines could accelerate the aircraft up to high speeds. Also, the top leadership wanted the aircraft to shatter the "sound barrier" immediately to eliminate the psychological barrier and quickly proceed to even higher performance speeds. To meet this second demand, the Air Corps ordered an aerial launch to save precious fuel for acceleration beyond the sound barrier rather than flight to altitude.

So what would the aircraft look like? What objects already exceeded the speed of sound? The scientists had their answer in a bullet. The first airplane to exceed Mach 1 was designed around the same outline of a 50-caliber machine gun bullet. With a pointy, sharp nose, smooth contour from front to back, and very small, thin wings to reduce drag, the Bell X-1 was born.

The Flight of the Bell X-1

On the morning of Tuesday, October 14, 1947, Captain Chuck Yeager climbed aboard the B-29 Superfortress that carried the Bell X-1 (see Figure 2-21) up to 20,000 feet. Captain Yeager, a decorated World War II *P-51* pilot and test pilot, climbed into the cockpit, wincing from the pain of broken ribs suffered when he was thrown from a horse days earlier. A celebrated tough guy, Yeager wasn't about to miss his chance.

At 10:26 AM and a speed of more than 250 mph, the B-29 released the small X-1, and Yeager fired up its rocket engine. Within minutes, the 6,000-pound thrust engines accelerated the low drag design up to and beyond the speed at which wind tunnel data could be collected (at that time). Flying into the unknown, Yeager did what any test pilot would do: He shut down two of the four engines and tested the aircraft's control systems. When he was satisfied that the aircraft could be flown, he reactivated the engines and accelerated through Mach 1 to a maximum speed of Mach 1.06, or 700 mph at 43,000 feet.

Captain Chuck Yeager had beaten the demons that many believed ruled the skies and demonstrated that supersonic flight was possible.

Figure 2-21 Bell X-1, World's First Supersonic Airplane.

Courtesy of NASA

Beyond Supersonic: How to Go Really Fast!

Our understanding of high-speed flight has been tremendously expanded in the years since the flight of the Bell X-1. The aerodynamics of high-speed flight has revealed a few basic guidelines for designing extremely fast aircraft:

▶ Use a sharp, pointy nose.

▶ Use extremely thin wings with sharp leading edges.

▶ Gently change the outline of the aircraft.

▶ Use an airfoil with a flat upper surface.

▶ Sweep the wings forward or aft, or switch to a delta wing.

▶ For hypersonic (Mach 5+), make everything blunt and rounded.

These last four elements require a little more explanation.

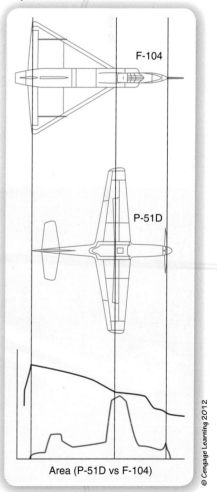

Figure 2-22 *The area rule determines cross-sectional area at various points in the structure of subsonic and supersonic aircraft.*

F-104

P-51D

Area (P-51D vs F-104)

© Cengage Learning 2012

The Area Rule

At subsonic speeds, air acts as if it were incompressible, easily flowing out of the way of an object such as an airplane moving through the air. When you get to higher speeds, this displacement becomes more difficult to do smoothly, which is why the sharp, pointy nose helps. It gets the air moving out of the way "gently" so it has time to step aside before even more air needs to flow around the airplane. However, no one had thought about what would happen when all that gently separating air was suddenly joined by a lot more air such as the air being shoved aside by the wing. Now there wasn't enough room for all the air to get out of the way fast enough, which produces a lot of drag. Someone realized that it was all about cross-sectional area. That is, if you slice through the airplane from left to right, how large will the airplane appear to the air at that point (see Figure 2-22)?

To give the air time to flow out of the way, this area had to change slowly and smoothly. This "area rule" meant that as there was more wing in the cross-section, there had to be less fuselage. Many aircraft gained a wasp-waisted appearance due to the area rule. How well does it work? When the YF-102 was first designed, it couldn't go supersonic (Figure 2-23a). However, after the area ruled was applied, drag dropped by 25%, and the aircraft was suddenly a supersonic fighter with the same engine (Figure 2-23b).

Figure 2-23 (a) The Convair YF-102 with cylindrical fuselage. (b) The Convair YF-102A with area rule tapered fuselage.

(a)

(b)

The Supercritical Airfoil

A supercritical airfoil is used to reduce the amount of drag induced by the shock wave produced when subsonic flight leads to supersonic airflow over the wing. A typical airfoil has a curved surface that causes the airflow over the top of the airfoil to accelerate to a high speed that exceeds Mach 1. A supercritical airfoil reduces the amount of acceleration using a flattened upper surface with more curvature near the trailing edge and with a much more rounded leading edge to reduce shock wave formation. Supercritical airfoils allow much higher speeds by reducing this **wave drag** and make the transition to supersonic flight less abrupt (see Figure 2-24).

Sweeping a Wing

How fast the air flows over a wing's airfoil shape is at least as fast as the forward speed of the airplane. After it interacts with the airfoil shape, it accelerates to even higher speeds. However, aeronautical engineers realized that the important speed wasn't the freestream airflow but the portion of speed that was at 90° to the leading edge. For a straight wing, these speeds are the same. But as the wing is swept backward (or forward), more of the airflow is spanwise. This makes a swept wing fly as though it was moving much slower than the airplane to which it is attached (see Figure 2-25).

Figure 2-24 F-104 Starfighter.

Figure 2-25 Skyrocket swept wing test plane with two F-86 chase planes.

Figure 2-25 Skyrocket swept wing test plane with two F-86 chase planes.

Courtesy of NASA Dryden Flight Research Center

Another advantage to wing sweep is that after the airplane goes supersonic, the sweep of the wings keeps them inside the shock wave created by the nose of the airplane.

The Blunt Rule

High speed was all about sharp edges and pointy features, but once you fly hypersonic, more than five times the speed of sound, everything changes. Below these speeds your main concern is drag reduction, but above these speeds the friction with the air will produce more than 300° of heating. This is enough to soften the aluminum materials used to construct a normal airplane. You can gain some relief by switching to stainless steel and titanium alloys, but eventually the heating becomes too great. At this point, most of the energy from the heating needs to go into the air rather than into the aircraft. By rounding all the edges and blunting the curves, you make a stronger shock wave that puts more of the energy into the air. For the Space Shuttle, which reenters the atmosphere at Mach 25, a low drag design would only heat the air about 3000° leaving most of the heat to flow into the craft (see Figure 2-26). With its rounded design, the actual shuttle heats the air around it to more than 14000° leaving much less of the energy remaining to enter and damage the craft! If you're going really fast, you need to go blunt!

Figure 2-26 Space Shuttle reentry conceptual artwork.

Courtesy of NASA Ames Research Center

THE RACE FOR SPACE: THE IMPORTANCE OF REACHING OUT OF THE ATMOSPHERE

Sending the first human out of the safety of Earth's atmosphere and into the blackness of space was an enormous scientific and engineering feat but also a major political move. It potentially held a much deeper meaning as a symbol of superior technological achievement and the ability to reach any part of the world in a matter of minutes.

Before Sputnik

"The Earth is the cradle of the mind, but we cannot live forever in a cradle." —(Konstantin E. Tsiolkovsky-Kaluga, 1911)

Rockets and space flight may seem like a relatively new concept, but the ideas and technology that shaped modern space flight have roots that extend back to the beginning of the 1900s. Although powder fuel or solid fuel rockets have been around since ancient China, they were considered as toys for amusement or as small ammunition for military purposes. However, solid fuel rockets continue through the modern world because they are fairly cheap to manufacture and are ideal for applications that require instant access such as surface-to-air missiles or air-to-air missiles. In most cases, a solid fuel rocket can be loaded ahead of time and placed in storage for a period of time. Other examples of solid fuel rockets can be seen in model rocketry, rocket flares, and the solid rocket boosters (SRBs) on the Space Shuttle. It wasn't until the beginning of the 20th century that we looked at alternatives that lead to the modern liquid fuel rockets.

In the same year the Wright brothers made their famous flight, a young Russian space scientist named Konstantin Tsiolkovsky published *The Exploration of Cosmic Space by Means of Reaction Devices*. Tsiolkovsky's book outlined the theory of spaceflight, weightlessness, and the basic equation for reaching space by means of a rocket, known as the Tsiolkovsky equation.

The original Tsiolkovsky equation:

$$\Delta v = v_e \ln \frac{m_i}{m_f}$$

To calculate final or initial mass for a given change in velocity, we need to rewrite the equation as

$$m_f = m_i e^{-\left(\frac{\Delta v}{C}\right)}$$

or

$$m_i = m_f e^{\left(\frac{\Delta v}{C}\right)}$$

where

- ▶ m_i = initial mass (kg or lbs).
- ▶ m_f = final mass (kg or lbs).
- ▶ v_e = exhaust velocity (m/s or ft/s).
- ▶ Δv = change in velocity produced (m/s or ft/s).
- ▶ C = effective exhaust velocity.
- ▶ e = the constant 2.71828.

Halfway around the globe at nearly the same time but completely separate from Tsiolkovsky, another scientist was developing his own ideas on space travel. While Tsiolkovsky developed basic rocket theory and outlines for space exploration concepts, Robert H. Goddard was an experimentalist, busy designing, building, and testing liquid fuel rockets. In 1914, Goddard received his first patents for a multistage liquid-propulsion rocket. Goddard's theories and mathematics were published in 1919 as "A Method of Reaching Extreme Altitudes," which along with Tsiolkovsky's theories, was the pioneering work that made modern rocketry possible. Of course, they didn't do it alone—like the Wright brothers and their bike shop, much of a rocket's success is based on ideas borrowed from other scientific fields. For example, Goddard's rocket was improved due to his use of Carl Gustaf Patrik de Laval's engine nozzle. Although originally designed for a steam turbine, de Laval's nozzle produced a 62% increase in efficiency of a rocket engine (see Figure 2-27).

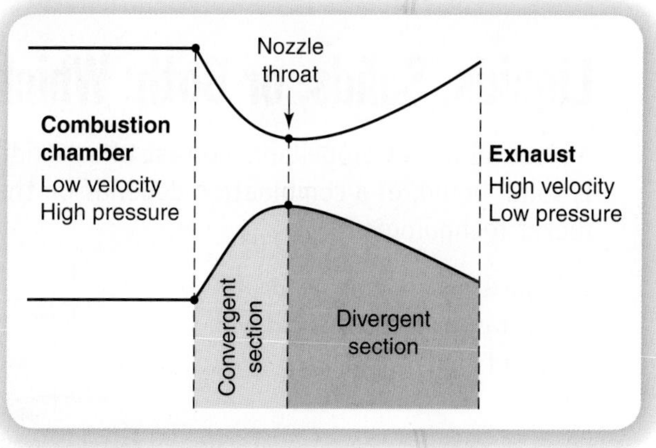

Goddard's launch of the first liquid-fueled rocket occurred on March 16, 1926 in a snowy field in Auburn, Massachusetts (see Figure 2-28). Goddard noted in his journal: "The first flight with a rocket using liquid propellants was made yesterday at Aunt Effie's farm" (from Clark University Archives, www.clarku.edu/research/archives/goddard/diary.cfm). With this modest understatement, the space race was on.

Figure 2-27 *The nozzle developed by de Laval generates 62% more thrust than its contemporary engines.*

Nozzle throat

Combustion chamber
Low velocity
High pressure

Exhaust
High velocity
Low pressure

Convergent section

Divergent section

© Cengage Learning 2012

Figure 2-28 Robert Goddard's liquid fueled rocket and launch stand.

Courtesy of NASA

The Race Is On: Fast Tracking the First Steps into Space

In the early days of World War II, German scientists and engineers played a pivotal role in the development of the rocket engine as a practical device. Primarily driven by political and military desires, Germany developed the first reliable rocket engine, the first rocket to reach the boundary of space at 100 km, the first rocket airplane (Me 163 Komet), and the V-2 rocket. In fact, Germany inevitably accelerated the space race to new heights for both the United States and the Soviet Union (now Russia).

Point of Interest

Liquids, Solids, or Both: Which Is the Fuel for Space?

Almost all rocket propellants consist of an oxidizer and a fuel. Whether the oxidizer and fuel combination is solid, liquid, or a combination depends on the purpose of the rocket. Figure 2-29 compares the existing rocket technologies.

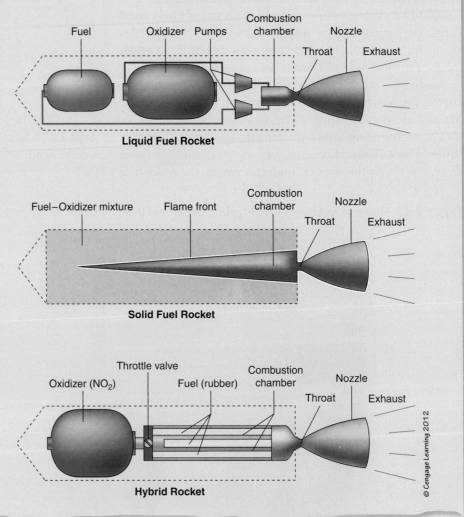

Figure 2-29 *Comparison chart of solid, liquid, and hybrid rocket engines.*

© Cengage Learning 2012

Just prior to World War II, Germany made its entry into the race. Hermann Oberth, with the help of his students at the Technical University of Berlin, conducted experiments and successfully tested the first German liquid-fueled rocket in 1929. One of his students, Wernher von Braun, would later lead Germany through the development of the V-2 rocket (also known as the A-4). The first successful launch of the V-2 occurred on October 3, 1942. From September 1944 through March 1945, nearly 3,200 V-2s were launched with 90% of the rockets impacting at or near London and Antwerp. Originally intended to be a long-range weapon, fortunately the technology of the time limited its range to 200 miles and prevented the V-2 from carrying a large payload capable of greater destruction.

Engineering Moments

Dissecting von Braun's V-2

Key technologies came together for the V-2's success. The rocket was used well past the end of the war as a test instrument because no other rocket could compete with it.

Propulsion:

▶ The liquid fuel bipropellant used liquid oxygen and alcohol mixture.

▶ Steam turbines powered by a pressured hydrogen peroxide mixture pumped the oxygen and alcohol mixture into the combustion chamber.

▶ The maximum burn time was 65-70 seconds with an operational range of 234 miles.

▶ 30 seconds after launch, the V-2 reached the speed of sound.

▶ The alcohol-water cooling mixture used in the V-2's engine had performance penalties but was necessary because high-quality metals for the nozzle throat were not available at the time. The temperature in the combustion chamber could reach 2,700°C.

Guidance Systems:

▶ Initially, the inaccuracy of the V-2 could vary from 4 to 11 miles off target. Towards the end of World War II, the guidance system was improved, but only a small percentage of the rockets used it.

▶ Guidance was based on gyroscope and accelerometer systems. After the gyroscopes were spun up before launch, their axis would remain constant no matter how the orientation of the rocket changed. This was combined with preprogrammed timers and simple electronics to achieve the target.

Stability:

▶ Development of the aerodynamic and control systems for the V-2 took hundreds of tests of models in wind tunnels, air drops, and powered flights. The missile had to be controlled from launch at near zero speed through supersonic speeds up to Mach 4. The final design incorporated four graphite rudders that controlled exhaust flow and four vanes on the fins. Gyroscopes controlled the vanes and rudders.

Putting it all together:

▶ Designs for the V-2 rocket were extensively tested (see Figure 2-30). Much of the testing was done with smaller models, but there was little experience in scaling up the engine and control technology to full size.

▶ From the initial design of the V-2 through the end of World War II, more than 65,000 changes were made to the design, much of it trial and error.

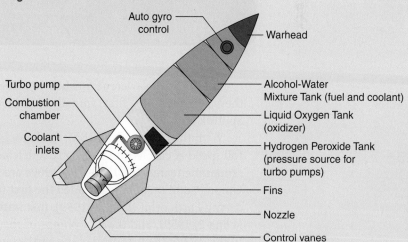

Figure 2-30 **The V-2 rockets of World War II.**

Auto gyro control — Warhead — Turbo pump — Combustion chamber — Coolant inlets — Alcohol-Water Mixture Tank (fuel and coolant) — Liquid Oxygen Tank (oxidizer) — Hydrogen Peroxide Tank (pressure source for turbo pumps) — Fins — Nozzle — Control vanes

© Cengage Learning 2012

As World War II ended, the Soviet Union and the United States rushed to capture as many of the rocket scientists and engineers as they could. At the foundation of the cold war, neither country wanted the other to lead in the space race. The United States captured most of the scientists while the Soviets captured most of the engineers and technicians. The U.S. cadre first went to Los Alamos, New Mexico, and then to Huntsville, Alabama. These scientists, including von Braun, continued improving the V-2 design but this time with a balanced emphasis on experimental and military value.

Meanwhile the Soviet Union expanded its rocket program development. Working on the initial theories and new developments by German scientists, both the Soviet and American rocket programs progressed at an alarming rate. By August of 1957, the Soviet Union had launched the first **intercontinental ballistic missile (ICBM)**. Just over six weeks later on October 4, the Soviets launched the first man-made satellite, named Sputnik, to orbit the earth (see Figure 2-31).

Figure 2-31 Sputnik, the first human-constructed object to orbit Earth.

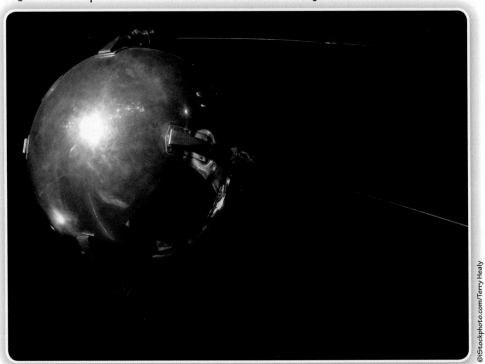

©iStockphoto.com/Terry Healy

But the Soviets weren't finished yet. Nearly one month to the day after Sputnik, they launched the first animal to orbit earth—a dog named Laika—aboard Sputnik II.

With the birth of the ICBM, the launch vehicle for Sputnik, the space race was full speed ahead. With the Germans defeated and their players divided among the remaining teams, the 1950s and 1960s are marked by one-upmanship. The Soviet Union's additional firsts included the first firing of a rocket in Earth's orbit and escaping Earth orbit (Luna 1, 1959), first impact of a man-made object on the Moon (Luna 2, 1959), and the first photos of the far side of the Moon exactly two years to the date of the first orbit of the earth (Luna 3, 1959). The United States managed a few firsts as well. The first pictures from space (Modified V-2, 1946), first animals in space (fruit flies aboard modified V-2, 1947), first solar-powered satellite (Vanguard 1, 1958), first communications satellite (Project SCORE, 1958), and the first weather satellite (Vanguard 2, 1959). That's five big hits for the Soviet Union and five hits for the United States. Even as the U.S. space program set new milestones during this middle stretch of the space race, they were under pressure due to the very public successes of the Soviet Union.

Through the rest of the decade, the United States sought to go above and beyond and achieve the impossible to gain dominance in the race for space. On January 31, 1961, Ham the Chimp successfully tested the same capsule American astronauts would use for spaceflight (see Figure 2-32). But on April 12, 1961, the Soviet Union hit a homerun when they sent Yuri Gagarin into orbital flight around Earth for the first manned spaceflight. In an attempt to catch up, on May 5, 1961, NASA launched Alan Shepard into space one month after Gagarin. Although Shepard passed well beyond the boundary of space, his flight lasted only 16 minutes, and he did not orbit Earth. It would take until February 20, 1962 before the United States could achieve a manned orbital flight with *Friendship 7* (*Mercury Atlas 6*) piloted by John Glenn. Almost from the start of the race, the Soviet Union set the pace. Feeling the political and military pressure, the United States made one of the boldest proclamations in the history of the space race.

A Determined National Effort

On May 25, 1961, President John F. Kennedy included the following statement in an address to the U.S. Congress on urgent national needs (Figure 2-33):

> ". . . if we are to win the battle that is now going on around the world between freedom and tyranny, the dramatic achievements in space which occurred in recent weeks should have made clear to us all, as did the Sputnik in 1957, the impact of this adventure on the minds of men everywhere, who are attempting to make a determination of which road they should take."

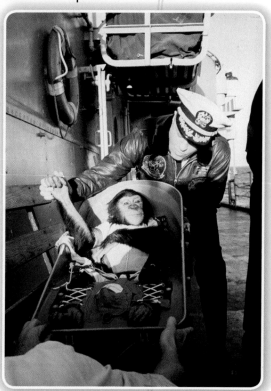

Figure 2-32 Ham the chimpanzee returns from outer space.

Courtesy of NASA

Figure 2-33 Kennedy's speech marks the beginning of the true space race.

Courtesy of NASA

Kennedy's address to Congress made it very clear that winning the space race was of vital importance, and the United States was not going to sit idly by. Kennedy continued in his speech by setting down the milestones in U.S. spaceflight that needed to be achieved, and laying out the path to be followed:

"First, I believe that this nation should commit itself to achieving the goal, before this decade is out, of landing a man on the Moon and returning him safely to the Earth. No single space project in this period will be more impressive to mankind or more important for the long-range exploration of space; and none will be so difficult or expensive to accomplish." (This script can be found at the John F Kennedy presidential library and museum website at www.jfklibrary.org/Historical+Resources/Archives/Reference+Desk/Speeches/JFK/Urgent+National+Needs+Page+4.htm.)

To put this statement in perspective, the United States had just launched its first astronaut into suborbital space. The Soviet Union had already sent several probes to the Moon, and was planning several more. On the political front, the cold war was escalating out of control and the failure of the Bay of Pigs invasion of Communist Cuba was nearly as devastating as being second to Gagarin's flight. President Kennedy, in an all-or-nothing statement, committed the American resources of NASA, scientists, engineers, and industry to the ultimate goal and timeline of putting a man on the Moon. The United States had only eight and a half years to develop a rocket system that could carry a lunar payload into orbit, transfer to a lunar orbit, land on the Moon, and return safely to Earth. The task seemed monumental and impossible but gave the United States a huge opportunity to recover lost ground. At the time of President Kennedy's speech, neither the Soviet Union nor the United States were anywhere near the capabilities of sending men to the Moon. The playing field was somewhat even despite the past shortcomings.

High Quality, Low Cost, and Quick Delivery—Pick Two

NASA needed this program to be high quality and delivered quickly. Cost was an enormous issue, but it would be far more devastating to the United States if the lunar missions failed or were delayed. Time was of the essence as the Soviet Union also raced to develop its lunar program. The initial cost estimates by NASA for the Apollo program reached nearly $35 billion. That's more than $187 billion in today's dollars! The Apollo program was so massive that it included more than 500 contractors, including most of the nation's major aircraft industries, to design, build, test, and refine the space hardware while NASA was dedicated to managing projects by overseeing quality control, assembly tasks, and timelines. As Mercury's one-man space capsule program continued its missions, work began developing the three-person Apollo hardware (see Figure 2-34).

There was an enormous gap between where NASA was with its technology and where it needed to be in less than eight years with a lunar landing. However, the most pressing issue facing NASA management was how to actually get to the Moon. Many theories were entertained, but they all revolved around three main methods:

▶ **Direct ascent.** This idea called for launching a large booster rocket directly to the Moon, landing a large craft on the lunar surface, and then sending all or some part of it back (see Figure 2-35). This idea was very costly and required an enormous amount of thrust, up to 40 million pounds, to get the craft out of Earth's orbit. This plan was scrapped early on as too impractical and expensive.

Figure 2-34 *Capsule size comparison.*

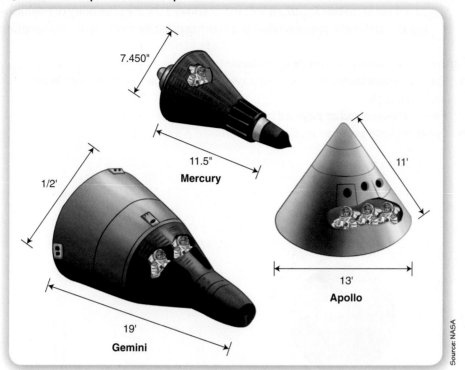

7.450"

11.5"
Mercury

1/2'

19'
Gemini

11'

13'
Apollo

Figure 2-35 *Direct ascent lunar landing.*

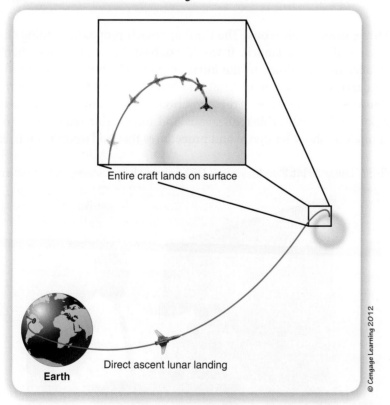

Entire craft lands on surface

Earth

Direct ascent lunar landing

▶ **Earth-orbit rendezvous.** With this approach, pieces of the lunar modules would be sent up into orbit above Earth separately. The modules would then rendezvous into a single system and be sent to the Moon (see Figure 2-36). The primary delivery system for this method would be the Saturn rocket, which delivered 7.5 million pounds of thrust and was already being developed. An added bonus to this approach was that it allowed for the possibility of establishing an Earth-orbit **space station** as a starting place for future space exploration.

Figure 2-36 Earth orbit rendezvous lunar landing
1. Two separate launches (fuel and payload/crew lunar vehicle) meet in orbit around Earth.
2. Fuel is transferred or payload boosts lunar rocket.
3. Lunar rockets proceed as direct ascent.

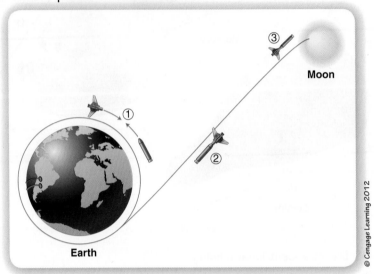

▶ **Lunar-orbit rendezvous.** The third approach proposed sending the entire lunar craft in one launch. It would establish lunar orbit, and then send a smaller module down to the lunar surface, which would then return and rendezvous with the main craft and head back to Earth (see Figure 2-37). This was the simplest of all three solutions but was still considered extremely risky. Both Earth-orbit rendezvous and lunar-orbit rendezvous made assumptions about hardware and procedures that had never been tried. Both

Figure 2-37 Lunar Orbit Rendezvous (LOR) solves the problems of weight and fuel.

required multiple spacecraft maneuvering and docking in space, and the possibility of astronauts having to leave the protected environment of the spacecraft to work in space. The lunar-orbit rendezvous had the additional danger of complex maneuvers occurring far from Earth.

While some systems such as the Saturn rocket could be built without knowing the exact method to get to the Moon, other systems could not proceed beyond concepts without knowing exactly how the lunar process would occur. It wasn't until November 7, 1962, a year and a half after President Kennedy's mandate, that the final decision for lunar-orbit rendezvous was announced.

Twins: The Gemini Missions

The technology gap between the Mercury capsules and the planned Apollo missions was vast but not impossible to overcome. NASA had been planning for a lunar landing as a long-term goal and as a problem to be solved as technology improved. The new timeline posed a problem, and the most logical solution gave birth to the Gemini program. The primary goal of Gemini was to perfect hardware and techniques for orbital maneuvering and rendezvous similar to what would be performed by the lunar missions. Other goals of the Gemini program included observations of human physiology in orbit for up to two weeks, extravehicular activities (spacewalks), and improved Earth reentry. The Gemini capsule was larger than the Mercury capsules with seating for two astronauts. The first of the manned Gemini missions started in March of 1965, halfway through the decade. In all, 10 missions and 16 astronauts were launched before the end of 1966 (see Figure 2-38).

Figure 2-38 *Gemini* timeline from March 1965 through November 1966.

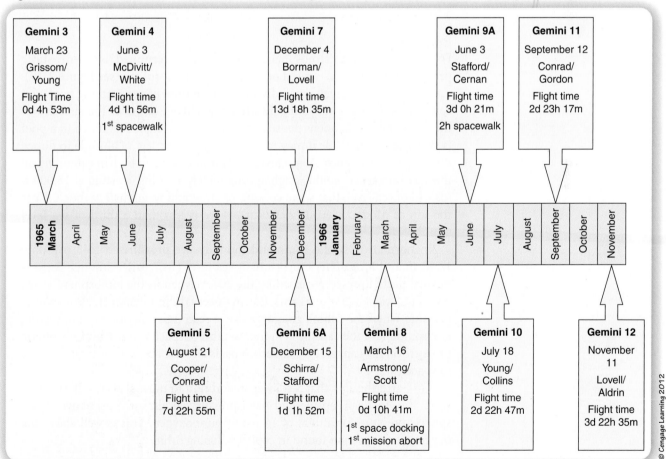

© Cengage Learning 2012

Gemini had its share of problems, from leaky fuel cells to shaky Titan II rocket systems, but the program was still considered a huge success, and the majority of the problems were eliminated for the Apollo program. The Gemini program accomplished all of its mission goals and completed 52 experiments that helped bridge the gap between the Mercury and Apollo programs.

Apollo: The Son of Zeus

Rough design for the *Apollo* spacecraft began almost immediately after President Kennedy's announcement in 1961. Engineers needed a plausible way to send astronauts to the Moon. Some initial design constraints included a command module that could keep three people alive for more than two weeks; a service module for holding fuel, maneuvering rockets, and other life-support systems; and the capability of using the command module for reentry into Earth's atmosphere. Development and testing of the *Apollo* spacecraft lasted nearly seven years from the delivery of the contract to North American Aviation (November 28, 1961) to the last test flight on October 22, 1968.

The Apollo program would have a traumatic interruption on January 27, 1967. During a manned "**plugs-out test**," *Apollo/Saturn 204*, later named *Apollo 1*, caught fire on the launch pad killing the three astronauts inside.

Virgil "Gus" Grissom, Ed White, and Roger Chaffee were performing what NASA considered to be a nonhazardous test in which the command module was pressurized with pure oxygen, similar to what would be encountered in orbit. However, one major oversight in a long line of other smaller mistakes culminated in the catastrophic failure.

Plugs-out test:

conditions recreated to mimic in all aspects a space launch, without actually launching the rocket. All other conditions set to match flight conditions as nearly as possible.

STEM Moment

Pressure

For the most part, atmospheric pressure is determined by hydrostatic pressure caused by the weight of the air in the atmosphere (see Figure 2-39). Just like the pressure felt when diving to the bottom of a swimming pool, you can feel the pressure on your ears when you change altitude. The air behaves the same way but is not as dense as the water so a couple feet down in a pool is very noticeable while it may take several hundred feet of altitude to notice the pressure difference. At sea level, a column of air 1 in by 1 in extending all the way into space would weigh approximately 14.7 lbs. Starting at 18,000 ft instead of sea level, that same column of air would weigh half as much. Near the top of Mount Everest (29,029 ft), there is less than a third of sea level atmosphere.

In normal orbit, the command modules are pressurized to 3 psi, only 1/5th that of atmospheric pressure at sea level (14.7 psi). Although 3 psi is normal partial pressure caused by the 20% oxygen in the atmosphere at sea level (14.7 psi $\times$ .20 = $\sim$3 psi), the engineers did not factor in atmospheric pressure when the cabin's gauge pressure read 3 psi. Because the operating environment of the capsule is in the vacuum of space where the atmospheric pressure on the capsule is near zero, 3 psi is also the absolute or total pressure in space.

On Earth though, that 3 psi was in addition to normal atmospheric pressure. The *Apollo 1* capsule had been pressurized almost 2 psi above atmospheric pressure for a total of 16 psi of pure oxygen. This is well above the oxygen concentration found in capsules during orbit.

Figure 2-39 Ninety percent of Earth's atmosphere is in the first two layers. Air molecules can still be found as far as 40,000 miles into space but are very far apart.

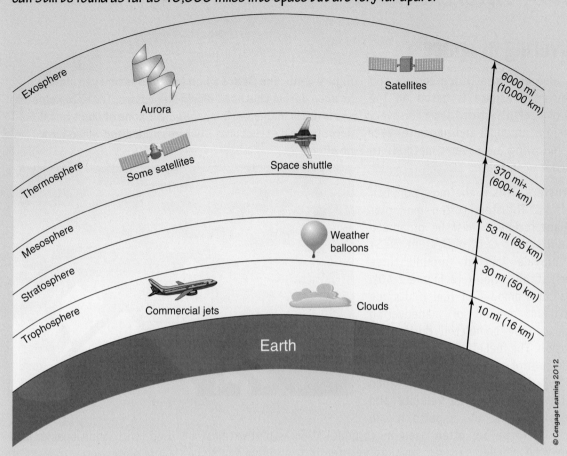

Exosphere

Aurora

Satellites

6000 mi
(10,000 km)

Thermosphere

Some satellites

Space shuttle

370 mi+
(600+ km)

Mesosphere

Weather
balloons

53 mi (85 km)

Stratosphere

30 mi (50 km)

Trophosphere

Commercial jets

Clouds

10 mi (16 km)

Earth

© Cengage Learning 2012

Air Mixture Concentrations and Space

The air we breathe is mainly comprised of two gases: nitrogen (78%) and oxygen (20%).

Normal sea level air pressure is 14.7 psi.

How much of that pressure is caused by the nitrogen gas? How much is caused by oxygen?

If Total Pressure (P_{tot}) = Atmospheric Pressure (P_{atm}) + Gauge Pressure (P_{gauge}), what is the Total Pressure of a capsule on a launch pad at sea level versus space if the gauge in the capsule reads 3 psi?

▶ Density of dry air at sea level: $68°F = 0.002377$ slug/ft³ ($20°C$, 1.204 kg/m³)

▶ Density of water at sea level: $68°F = 1.936$ slugs/ft³ ($20°C$, 998.2071 kg/m³)

▶ Actual percentage of oxygen in air: 20.946%

Point of Interest

Maneuvering in Space

The basic idea behind attitude control and fine maneuvers in space is based on the principles of Newton's third law: For every action, there is an equal and opposite reaction. The following are several methods to maintain control in space:

► **Thrusters.** This is the most common method. Thrusters (often monopropellant rockets), must be organized as a Reaction Control System (RCS) to provide triaxial (around three axes) stabilization (see Figure 2-40). The total amount of fuel limits the maximum amount of maneuvering that can be completed during the mission. The fuel efficiency of an attitude control system is determined by its ISP (in-space propulsion), which is the rocket's exhaust velocity, and the smallest torque impulse it can provide. These are often used in conjunction with smaller systems such as vernier thrusters that accelerate ionized gases electrically using power from solar cells.

► **Spin stabilization.** The entire space vehicle itself can be spun up to stabilize the orientation of a single vehicle axis. This method is widely used to stabilize the final stage of a launch vehicle. The entire spacecraft and an attached solid rocket motor are spun up about the rocket's thrust axis.

► **Momentum wheels.** These are electric motor-driven rotors made to spin in the direction opposite to that required to reorient the vehicle. Because momentum wheels make up a small fraction of the spacecraft's mass and are computer controlled, they give precise control. Momentum wheels are generally suspended on magnetic bearings to avoid bearing friction and breakdown problems. To maintain orientation in three-dimensional space, a minimum

Figure 2-40 The Space Shuttle RCS system can clearly be seen during this heat shield inspection. The RCS helps the shuttle maneuver in space where none of the typical aerodynamic structures such as rudder and wings have an effect.

Courtesy of NASA GRC

of two must be used, with additional units providing single-failure protection.

► **Control Moment Gyros (CMG).** These are rotors spun at constant speed, mounted on gimbals to provide attitude control. Although a CMG provides control about the two axes orthogonal (at 90°) to the gyro spin axis, triaxial control still requires two units. A CMG is a bit more expensive in terms of cost and mass because gimbals and their drive motors must be provided. A major drawback is the additional complexity, which increases the number of failure points. For this reason, the International Space Station (ISS) uses a set of four CMGs to provide dual failure tolerance.

► **Solar sails.** Solar sails that reflect light and produce a very small amount of thrust may be used to make attitude adjustments. This approach is fuel efficient for long-duration missions that need only minor adjustments.

The super-saturated pure oxygen environment (combined with 34 sq ft of Velcro and 70 lbs of other flammable materials) was ripe for an ignition. Although the actual cause of the fire was never determined, it is suspected that a short in a wire near the command module pilot or a strong static discharge may have initiated the fire. *Apollo 1* was a tragic event, but it led to many design changes and new testing procedures. Some of the design changes included an outwardly opening hatch, removal of flammable materials, wires covered in protective insulation, and Beta cloth, which is nonflammable, melt-resistant replacements for nylon space suits. Had this tragedy occurred in flight on the actual mission in February of 1967, not only would the three astronauts and the *Apollo* spacecraft have been lost but also the accident might have been the end of or seriously delayed the achievement of Kennedy's mandate. The *Apollo 1* incident delayed the program by almost two years, but the changes made in procedures and the spacecraft led to the first successful manned launch of *Apollo 7*'s Earth-only orbit on October 11, 1968.

NASA engineers and managers were faced with yet another challenge. Delays were keeping the Lunar Module from completion, and when it finally shipped to Cape Canaveral, several defects scrubbed its flight-ready status. In an effort to keep schedule and thwart a possible Soviet circumlunar flight, NASA created the *C-Prime mission* for *Apollo 8*. This mission would take the Command Service Module and its three astronauts all the way to the Moon for 10 lunar orbits but would not land on the lunar surface. On December 21, 1968, *Apollo 8* left Earth and headed for the Moon. Almost 70 hours after launch, *Apollo 8* entered lunar orbit. In addition to testing the Command Service Module's capabilities and several live TV broadcasts, mission objectives included high-resolution photos of proposed landing sites for future Apollo missions. The *Apollo 8* mission lasted 6 days and 4 hours.

One Small Step, But Miles to Get There

Apollo 9 and *Apollo 10* were baby steps to the Moon. *Apollo 9* did not actually go to the Moon. From March 3 to March 13, 1969, it stayed around Earth but performed 152 orbits. This was the first time the complete Apollo Lunar system, Command/Service Module, and Lunar Module were actually in space.

During the orbits, the crew demonstrated that the complete system worked as intended by completing all steps involved in the Apollo mission without actually going to the Moon.

For the final dress rehearsal, *Apollo 10* would travel to the Moon and do everything except land on the Moon. Lunar Module Pilot Eugene Cernan and Commander Thomas Stafford would bring the Lunar Module 50,000 ft off the Moon's surface but not land it. A couple years later, Eugene Cernan would be the last man on the Moon with *Apollo 17* (see Figure 2-41).

Figure 2-41 *Apollo 17* astronaut Harrison (Jack) Schmitt pauses for a moment to pose with the American flag and rover.

Courtesy of NASA Johnson Space Center

Going to the Moon

Following are the basic challenges NASA engineers faced in the quest to go to the Moon.

▶ Insertion into orbit

▶ Separation of vehicles from launch formation

▶ Maneuvers to dock the Command/Service Module to the Lunar Module

▶ Service module burn for escape velocity

▶ Various attitude maneuvers to test systems

▶ Service module burn for lunar orbit

▶ Lunar Module evaluation and crew transfer

▶ Command/Service Module and Lunar Module separation

▶ Lunar Module decent engine burn

▶ Various maneuvers for lunar surface landing

▶ Lunar Module ascent burn

▶ Command/Service Module and Lunar Module rendezvous and docking

▶ Command/Service Module lunar orbit escape burn

▶ Jettison Lunar Module from Command/Service Module

▶ Command Module and Service Module separation

▶ Command Module reentry

The preparations and dress rehearsals had gone as planned; two months after the *Apollo 10* mission; NASA was ready for the Moon landing. *Apollo 11* was prepared for lift off. At 9:32 AM on July16, 1969, from launch pad 39-A, *Apollo 11* and its crew, Neil Armstrong, Edwin E. Aldrin, Jr, and Michael Collins, left for the Moon. Four days later on July 20, Neil Armstrong would make his first step on the Moon at 10:56 PM, EDT (see Figure 2-42), and say, "That's one small step for man, one giant leap for mankind."

There were six more Apollo missions, although the last three missions, 18, 19, and 20, were cancelled due to budget issues and the new Skylab project. Of the six remaining Apollo missions, only five actually made it to the Moon. *Apollo 13*, during its travel to the Moon suffered a rupture in a service module oxygen tank. The mission was classified as a "successful failure" due to the effort of ground crew and engineers to bring the crew safely home.

Figure 2-42 *Apollo 11 astronaut Neil Armstrong leaves behind a footprint that could last for generations.*

Courtesy of NASA Johnson Space Center

The remaining Apollo missions performed several <mark>extravehicular activities</mark> (EVAs) on the lunar surface. A large part of the EVA missions involved obtaining soil and rock samples from the lunar surface. In an effort to cover more ground than the astronauts were able to cover by walking or rather hopping on the Moon's surface, the <mark>lunar roving vehicle</mark> (LRV) was brought up on *Apollo 15* (see Figure 2-43). The LRV was built by Boeing Aerospace but had to survive the harsh conditions on the Moon. No vehicle like the LRV had ever been built previously. To get the LRV to the Moon, Boeing designed it to fit in a small storage space under the lunar lander (see Figure 2-44). When on the Moon, the LRV would transform from a small pie-shaped package into a four-wheel vehicle with little help from the astronauts. Unfolded, the LRV was about 10 ft long and 6 ft wide but only weighed about 460 lbs—slightly smaller than a Mazda Miata convertible, which weighs more than 2,000 lbs. The LRV was capable of carrying two astronauts, their gear, and lunar samples for a total payload of a little more than 1,000 lbs—more than twice the vehicle weight. Each wheel was individually powered by a quarter horsepower electric motor for a top speed of about 8 mph, which was more than enough for the low-gravity, rough terrain of the Moon.

In all, 10 LRVs were made, but only three made it to the lunar surface. The others were used for ground testing and training here on Earth. The LRV greatly increased the distance covered on the lunar surface in addition to aiding in gathering samples of interest (Figure 2-45).

Figure 2-43 The lunar rover increased the exploration area and provided a great platform for equipment for the lunar astronauts.

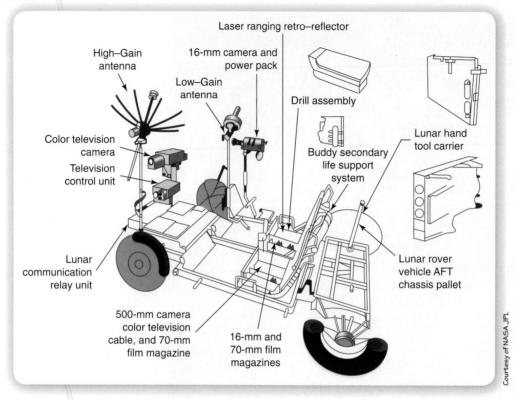

Figure 2-44 Lunar rover being loaded onto the Lunar Excursion Module (LEM).

Figure 2-45 Lunar rock sample from *Apollo 11, Sample # 10046.*

The 842 lbs (382 Kg) of lunar samples collected by missions that landed on the Moon did not go to waste (an additional 1 lb of lunar samples were collected from Soviet Luna probes). All of the lunar samples collected by the six Apollo missions are considered priceless and are kept in special containers to keep atmospheric contamination from ruining the samples. The samples can be found in museums such as the National Air and Space Museum and the Kennedy Space Center.

Several small samples were mounted and presented to national governments. NASA estimates that close to 77% of the original samples are still in their pristine condition safe in the vault at Johnson Space Center.

Point of Interest

The Soviet Lunar Program

The race was not a one-man show when the United States set the bar for the Moon. The Soviet space program made several attempts for the Moon but was plagued by engineering and technical problems. By 1963, both sides had the preliminary designs of the launch platforms to get the lunar hardware into orbit. In 1967, catastrophe endangered both lunar programs with the fire of *Apollo 1* in the United States and the crash landing of the Soyuz spacecraft in the Soviet Union. The Soviet program was slowly making progress but encountered more setbacks and fatalities in 1968. By the end of that year, the United States had sent *Apollo 8* around the Moon while four attempts of the unmanned Soviet Zond circumlunar missions failed. In 1969, the U.S. space program was on the lunar surface twice, but on the Soviet side, there was another failed unmanned Zond mission followed by the only successful Zond mission (*Zond 7*) near the end of the year. This success, however, came too late as the final Zond mission in 1970 was only partly successful, and the Soviet lunar program was canceled and resources were reassigned to Space Station missions by 1974.

Lunar Lunacy

On December 14, 1972, Eugene Cernan of *Apollo 17* stepped off the lunar surface and back into the Lunar Module, effectively ending man's adventures on the Moon. Several decades would pass before humans considered returning to the Moon. At least four countries—the United States, China, Russia, and Japan—are developing manned missions to the Moon scheduled for sometime before 2020. China was the third country to place a man into space (2003) and is aggressively pursuing its lunar goals, almost at the pace of the early Apollo timeline. So why the big push to return to the Moon? Countries are motivated by national pride, the potential as a launch platform for future Mars missions, and, most importantly, the potential of harvesting lunar natural resources such as frozen water and other raw elements vital to a permanent base on the Moon.

AUTONOMOUS SPACE EXPLORATION

Manned spaceflight is a complex task. Humans are very fragile above the Earth's surface and beyond its comfortable atmosphere. We need air to breath, food to eat, and time to sleep and rest. Any space flight or travel for humans requires all of these things at bare minimum. As we look at Mars and beyond as points of interest in space for human exploration, we must consider these details. The alternative is sending a probe or robot to do the work of exploration.

Introduction

A probe can be much smaller because it does not have to carry the life-support systems or the living space for a human and can still be built to endure the harsh environment of space. Life-support systems add cost to any spacecraft while consuming large portions of the weight and volume the launch vehicle can lift. Additional dangers include exposure to radiation and greater impact of physiological effects due to weightlessness. Because they can be smaller and less complicated, the overall cost of an unmanned space probe is reduced. Long has the debate continued over the need for manned spaceflight. Why send a human when we have the technology that can near duplicate the efforts of a human? The most successful space programs will never completely be one or the other but instead complimentary manned and unmanned programs. Robots will explore and gather initial data, and humans will follow behind to interpret and use the data.

The Price of Human Achievement

The impact of Yuri Gagarin's first spaceflight and the famous first words of Neil Armstrong when he stepped on the Moon were huge emotional moments for humanity. Naturally, it is hard for us to get as emotional over a probe or robot that has done significantly greater tasks even though they are devices humanity built. The emotional impact is equally similar when tragedy follows manned spaceflight. Of all manned space missions, 90% have been successful. There are only four events in which fatalities occurred during an actual mission and none in actual space: Soyuz 1 (April 23, 1967), Soyuz 11 (June 6-29, 1971), Challenger (STS-51L) (January 28, 1986), and Columbia (STS-107) (February 1, 2003). In total, 18 of the more than 450 people that have been in space have perished during a mission. Keep in mind that the number of people actually in space compared to the number of actual missions is low because many astronauts have been to space more than one time. It is difficult for the scientific community to put a value on human life, but the astronauts know spaceflight is risky. The success rate of space probes is marred by many more failures than manned missions. Engineers are not sacrificing human lives so a probe failure is more acceptable, although they always strive for success due to the time limitations and the cost of each probe.

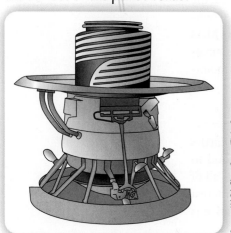

Figure 2-46 *Venera 9 probe sent to the surface of the planet Venus.*

Where No Human Has Gone Before

Probes have proven to be an invaluable resource for space exploration. Some of the most famous probes are in places that humans can only dream of going. The Soviet space program sent several probes (Luna series) to the Moon 10 years before humans first set foot there. From 1961 through 1983, the Soviet Union sent 16 probes to Venus with all but the first three successfully completing their missions. In 1970, *Venera 7* was the first probe to land and transmit data from the surface of another planet (see Figure 2-46). The probe sent back crucial information about the atmosphere of Venus such as the pressure (90 atm) and temperature (465°C). For perspective, consider that a typical Earth day at the beach is 1 atm and 28°C. The harsh conditions on the surface of Venus quickly melted components of the probes, allowing a total of less than 30 minutes of data to be broadcast between two successful landers.

UNMANNED AERIAL VEHICLES (UAVs)

Many have the false notion that Unmanned Aerial Vehicles or Uninhabited Aerial Vehicles (UAVs) are a new invention. The UAV in some form or fashion has been around since 1916 when the Hewitt-Sperry Automatic Airplane was developed. The automatic control equipment was originally tested on a Curtiss N-9 seaplane, and then transitioned to a wartime effort in the form of the Curtiss-Sperry Flying Bomb. The flying bomb airframe was the first to be tested in an open-air wind tunnel via the top of a car. After optimizing the design, the flying bomb was redesigned to launch from the top of a moving vehicle. This method worked on March 6, 1918, and the flying bomb maintained steady and controlled flight for the preprogrammed 1000-ft flight. In the end, this was the only successful flight, and the project was abandoned. Although the project ended in failure when all the test airframes were destroyed in crashes, several key engineering breakthroughs occurred.

The technology developed for guiding the flying bomb was based on gyroscopic control, which developed into the basis for autopilot control in modern aircraft. The more promising technology that led to the success of the modern UAV rested in radio control (RC). The Hewitt-Sperry Automatic Airplane was originally intended to use a combination of gyroscopic and radio control. Nikola Tesla is credited with using the first radio-controlled device in 1898 on a small boat, but the first radio-controlled aircraft would go to Archibald Low. Low developed, with the help of the Royal Flying Corps, the "aerial target" (AT), which, for strategic purposes, was a misnomer because the AT was really a flying bomb designed to carry a warhead. On March 21, 1917, the radio-controlled aircraft was born with a successful test flight. Unfortunately, Low's ideas were years ahead of their time and would not reach their full potential until the late 1930s and World War II. Several RC projects from the Axis and the Allies were developed primarily using electromechanical devices to operate relays to control actuators. This system was bulky and prone to malfunction but was used until the 1960s when the introduction of solid state control systems improved radio control. The basic mechanism was still the same: A signal from the ground was received by the radio control receiver, which transmitted to a relay that moved a control actuator and a control surface. Though modern RC equipment has much smaller components, providing more control channels, elements of both radio and gyroscopic control remain in today's autonomous vehicles with visual and telemetry data feeds back to the pilot on the ground. Since Low's first days with the RC aircraft, it was primarily directed to military use with the remote craft being destroyed in the form of a flying target or remote bomb.

Today, we define a UAV as a remote-controlled or fully autonomous vehicle that can be reused many times or is not intended to be destroyed. Today's UAV comes in many shapes and sizes, all with distinct purposes. Some uninhabited aircraft work independently or are part of a system known as an Uninhabited Aerial Vehicle System or just Uninhabited Aerial System. The attack-capable Predator, which has flown many hours over the Middle East and Bosnia, includes up to four aircraft with sensors, a ground control station, satellite link, and approximately 55 personnel, and is deployable anywhere in the world. Attack-capable UAVs such as the Predator are also known as Uninhabited Combat Aerial Vehicles (UCAV).

The idea of using UAVs for reconnaissance was a product of the cold war, in which both the United States and the Soviet Union developed and deployed long-range reconnaissance drones in secrecy. A beautiful example of a modern

Actuators:

a device designed to create a mechanical force that causes something to move. Actuators can open or close valves or move a control surface to a position.

Figure 2-47 *Global Hawk represents a modern Unstaffed Aerial Vehicle (UAV).*

Courtesy of NASA Dryden Flight Research Center

reconnaissance UAV is the Global Hawk manufactured by Northrop Grumman (see Figure 2-47). The Global Hawk has a huge wingspan and high aspect ratio similar to what is found on sport gliders. The main purpose is to give the Global Hawk endurance at high altitude flight where it can maintain position for up to 42 hours.

Future of UAV

Following the success of the Predator UAS, UCAVs and UAVs continue to grow in many capabilities. Some experts believe the F-35 Joint Strike Fighter (JSF) will be one of the last manned fighters built, being replaced by smaller, cheaper, and more maneuverable UAVs. Remember, current manned fighter aircraft have much greater power and maneuverability than can be endured by the pilot but are governed by computer control to not exceed those limits. Additionally, cabin space, life support, and safety systems add precious weight to the aircraft. Moving the pilot to ground-based systems addresses those issues and takes the pilot out of harm's way. At the same time, this raises other issues involved with UAVs, especially those designed for attacking targets.

Leaving the pilot on the ground to watch through a camera could impact their decisions, not allowing the pilot to have full understanding of a situation or target. And still more concerning will be attack decisions completely made by a computer. That scenario may still be a long way off, but the evolution is there.

Whereas some UAVs and UCAVs are designed for stealth capabilities, other UAVs are designed for portability. The Army and Marines use a Micro Unmanned Aerial Vehicle (MAV), called the Raven, which is small enough to fit in two packs that weigh a combined total of 10 lbs (4.3Kg) and can be launched with a two-man crew (see Figure 2-48). The Raven provides Special Forces teams with day or night surveillance of a target area to allow evaluation of possible hazards before entering an area. MAVs, like the Raven, which can be directly flown by joystick or a GPS plotted course, have been employed by the military since 1989.

Palm-size micro-UAVs are being developed for surveillance (see Figure 2-49). The Defense Advanced Research Projects Agency (DARPA) is supporting a

Figure 2-48 A Honeywell Micro Aerial Vehicle (MAV).

Figure 2-49 The TACMAV is stored in a tube when not in use and can either be flown manually or autonomously via portable computer.

program that will develop several types of micro-UAVS. The idea is that several inexpensive micro UAVs, similar to AeroVironment's prototype Black Widow, would work as a team to observe an area.

Some of the more complicated UAV systems are based on hovering or rotary wing aircraft (helicopters). Helicopters have the unique advantage of maintaining stationary flight, which can be very useful for a small UAV to monitor a target area. However, this poses two problems: A stationary target is an easy target, and rotary wing aircraft can be very loud, making stealth difficult.

The Association for Unmanned Vehicle Systems International hosts competitions for universities to develop cheaper and more reliable UAVs. The contest parameters require teams to do a complete autonomous mission, including autonomous takeoff, GPS navigation, automatic target identification, and autonomous landing. Every few years, the Association changes the contest to keep challenges competitive based on current technology (see Figure 2-50).

Figure 2-50 The University of Texas at Arlington's Autonomous Vehicle Labs (AVL) has won several competitions and categories using custom electronics and off-the-shelf RC aircraft components.

Image courtesy of Universty of Texas Arlington

Source: Compliments of Dr. Arthur Reyes

ON THE HORIZON

Projects such as the various X-prizes for nongovernmental development of aerospace goals will help boost the field in new directions. The methods for getting to space are changing along with the reasons for going. For many years, the only way to get to outer space was by rocket or booster rockets. Engineers are now looking back to old designs for inspiration while thinking creatively for new innovations.

Going Retro: The Return to Space—Capsule Style

The Space Shuttle is the work horse of NASA. Designed to be an economically reusable space transport, the shuttle fleet has performed hundreds of missions and carried tons of hardware into space. Development of the shuttle began in the early 1970s, and its maiden space flight occurred in 1981. Although the basic shape of the shuttle has changed little since the development, the electronics and safety systems have gone through many overhauls as technology advanced.

The shuttle system was not as economically viable as initially estimated. With NASA's continually tightening budget, the unfortunate accidents of *Challenger* in 1986 and of *Columbia* in 2003 increased safety protocols, and inspections have pushed the average cost of a shuttle launch to well over $450 million. The build cost to replace *Challenger* with *Endeavour* was in excess of $1.7 billion in 1991. The shuttle system is an extremely reliable system despite the two accidents, but it is an expensive and aging system. In 2004, NASA announced they would retire the Space Shuttle to be replaced by the Orion rocket system (see Figure 2-51). Developed with expandability in mind, the Orion system will be used for low-Earth orbit, lunar, and possibly Mars missions. The initial designs of the Orion Crew and Service Module look strikingly similar to the Apollo Command Module, with some key differences.

Figure 2-51 NASA *Orion Lunar Command Module* orbiting the Moon (artist's conception).

Courtesy of NASA

The Orion Command Module will hold a larger crew, has much more computing power (most modern calculators have more computing power than the Apollo systems), and will employ the ability to refurbish and reuse the Command Module for up to 10 missions. In an effort to save development costs, the Orion system will also be fitted to use one of the Space Shuttle solid rocket boosters (SRB) for a launch stage. The total cost through 2025 for the Orion project (Project Constellation), including inflation, is estimated at $217 billion. This figure includes several years of development, testing, and manned lunar missions. This compares to the Apollo budget, however, Project Constellation will span at least 10 years more than Apollo.

Top Floor: Exosphere, Van Allen Belts, and Geosynchronous Orbit

One of the more bold designs involves a space elevator to take cargo and personnel into space. This idea, originally conceived by Tsiolkovsky and later explored by Arthur C. Clark, has finally gained serious attention as technology gets closer to making the space elevator a possibility. Many obstacles both in design and physics still stand in the way of making this proposal a reality, but the most promising technologies for a lightweight tether rest in carbon nano-tubes. The act of an elevator moving up and down the tether is subject to the Coriolis effect caused by Earth's rotation, which will cause issues with the tether and could cause it to break if it is not strong enough to handle the load. Most likely the 1/6th gravity and lack of atmosphere of the Moon would make it a more likely practical place to try a space elevator to get lunar resources into space.

Hybrids in Space

Until recently, there were two main types of rocket engines: solid fuel, which uses a mixture of a solid fuel, oxidizer, and catalyst, or liquid fuel, which uses a combination of liquid hydrogen and liquid oxygen. A solid-fuel rocket is much simpler than a liquid-fuel rocket and provides relatively high thrust for the cost. As a consequence, however, solid-fueled rocket exhaust can be very toxic to the environment and cannot be shut off once ignited. Liquid-fuel rockets have a much higher cost-for-thrust ratio due to the engine complexity required for mixing hydrogen and oxygen.

The advantages of a liquid-fueled system allow for a much cleaner exhaust by-product (steam) and the ability to throttle or shut off the thrust completely. It's only logical that the next evolution in rocket engines is a combination of both systems in a hybrid rocket engine (see Figure 2-52). Hybrid engines use a fuel such as ABS plastic, synthetic rubber or paraffin wax, and a gaseous or liquid oxidizer such as oxygen or nitrous oxide. Hybrid engines are not impervious to failure. They can

Hybrid:

combining two or more technologies or components to perform essentially the same task.

Figure 2-52 SpaceShipOne's Hybrid Rocket developed by SpaceDev is mounted to the Nitrogen Oxide tank prior to test firing.

image courtesy of SpaceDev

have issues with pressure vessels exploding if not properly constructed and controlled. Hybrid engines are low cost and have a very low explosive equivalent because usually both fuel and oxidizer are inert until combined, compared to a solid fuel rocket, which is usually 100% explosive.

Moon Patrol: Establishing a Permanent Lunar Station

The proposed timeline for the lunar future includes a permanent manned base station by 2024. With such a distant goal and rapidly developing technology, the lunar base design is wide open. Several basic constraints exist despite the final design selected. For example, the ideal location for the Moon base is at one of the lunar poles where constant sunshine for solar power is abundant. The base will need to eventually become self sufficient, mining the lunar resources for fuel, oxygen, and other materials. Future plans for the lunar base include becoming the Mars mission refueling station and providing access to manufacturing technologies in the perfect vacuum and low gravity of the Moon. NASA's Institute for Advanced Concepts has awarded grant money for companies to investigate a lunar elevator to work in conjunction with the lunar base. The elevator would not be suitable for human transport because it would have to move at a slow speed that could take weeks to complete transfer to orbital heights.

Challenges for Manned Mars Missions

NASA and the Chinese/Russian governments are looking at Mars for long-term plans. It may take a combined effort from both agencies to make it a reality though. Mars has a longer orbit than Earth, which means the **least energy missions** to Mars occur every 26 months when Earth, Mars, and the Sun nearly line up. Also, every year, the orbits of the two planets around the Sun vary to some degree, making the calculations for engineers very difficult. A Mars year is approximately 687 Earth days and therefore is in a much longer orbit around the Sun. The orbit brings Mars to within 55,700,000 miles of Earth at closest approach, but the distance increases to more than 401,000,000 miles when they are on opposite sides of the Sun. When launched at the closest distance, payloads still take 3 months to reach Mars. At the maximum distances, it takes radio communications 23 minutes to travel one way from Mars to Earth, and the Sun creates enormous amounts of radio interference, making real-time communications impossible. The most productive and successful Mars rovers have been the Mars Exploration robots Spirit and Opportunity, which were launched June 10 and July 7, 2003. The rovers initial mission timeline was to last only 90 days, but the rovers have maintained communications and function and operations have been extended until 2009. Both rovers are over 4 years old and have traversed several kilometers on the harsh Martian surface (see Figure 2-53).

Out of the 31 attempted missions to Mars, only 5 have successfully sent data back to Earth. Until this success rate can be improved, Mars exploration will remain the job of robots while manned missions to Mars remain a dream for future generations.

Space Tourism

Space tourism allows a person to pay for the experience of traveling into space for personal satisfaction. Space tourists are usually people who spend insane amounts

Courtesy of NASA Jet Propulsion Laboratory

of money for a relatively short but very unique experience. The first space tourist was Dennis Tito, an American businessman, who paid Space Adventures, Ltd. $20 million for a 10-day trip to the International Space Station. But these tourists do not always have to pay nor are they called tourists. The Russian space program and NASA have sent several nongovernment "payload specialists" who were not given the same training or stringent physicals as regular astronauts. Several payload specialists had missions that were public relations based and were not scientific or mission critical such as congressional sponsor John Glenn. Barbara Morgan who was the backup for teacher Christa McAuliffe later became a fully qualified astronaut. For the Russian space station MIR, a substantial fee allowed Japanese reporter Toyohiro Akiyama one week on the space station. The following year, England selected chemist Helen Sharman from a pool of applicants to go to MIR, making her the first Briton in space. Although neither paid their own ticket, the foundation was set—space tourism was big money. Since 2003, space tourists have been referred to as "spaceflight participants." As the commercialization of space expands and governmental space agencies continue to search for funding, the ISS remains the premier space participant destination; however, options are on the horizon.

Sir Richard Branson and Virgin Galactic are selling seats aboard *Spaceship Two*. Spaceship Two, the big sister to the *Spaceship One* winner of the Ansari X-Prize, will accommodate six passengers and two crew members to nearly 65 miles above Earth's surface. The flight will last approximately two and a half hours and cost $200,000. For the price of the ticket, the spaceflight participant gets three days of training and a momentarily spectacular view.

Establishing a colony on the Moon is still many decades away but that is not delaying private companies from selling Moon tours. Future tourism plans for the Moon include lunar flybys, which could last anywhere from 10–21 days, and long-term plans of a lunar hotel. Much of that is speculation at this point because it will still be at least 2018 before humans step on the Moon again.

Careers in Aerospace Engineering

SPREADING HER WINGS

"When I was a kid, I thought the work that NASA was doing was really cool—the space shuttle, the missions to Mars."

That's Angela Schroeder, now a senior engineer with the Cessna Aircraft Company, talking about what inspired her to become an engineer. "I was always good at math and science," she says, "and I liked figuring out how things worked, but I didn't know how that connected with something I might actually do for a living."

She eventually figured that one out as well.

Angela Schroeder

© Cengage Learning 2012

On the Job

Today, Schroeder works on advanced-design programs for Cessna, which has been putting planes in the air for more than 80 years. "I do structural analysis to optimize size and weight," she says. "If you want a plane that's faster and more fuel efficient as well as able to carry more passengers and more cargo, you need an airframe that can handle all that—strong, but not too strong, because then the plane would be too heavy, and that's a waste of resources."

Schroeder occasionally heads over to the production line or checks out the latest experimental aircraft as it's taking shape, but like so many engineers, she spends a lot of her time doing simulations on a computer. That's where she worked on a problem that has plagued airplanes since the days of the Wright brothers: bird strikes. "The FAA sets guidelines on which part of the airplane needs to withstand the impacts," Schroeder says, "but the design of that structure is different for every airplane, and you're continually having to refine your analysis."

Education

It wasn't until the summer before college that Schroeder realized she could apply her aptitude for math and science to a career in engineering. "I was on my lunch break at work one day," she says, "and something came on the radio about NASA. I was saying how cool I thought it was, and my co-workers all said, 'You could do that, too.' It had never occurred to me before."

A first-semester intro-to-engineering course at Washington University in St. Louis solidified Schroeder's thinking. "Each week, we'd cover a different branch of engineering, with a different project assigned for each one," she says. "Mechanical engineering seemed like the best fit—the one that best spoke to my creative side."

Schroeder followed up her BS in mechanical engineering with an MS in aerospace engineering from the University of Maryland. Actually, it was a joint program with the NASA Langley Research Center, and it led to one of Schroeder's most rewarding assignments. "I was on a team that processed data from the Mars Reconnaissance Orbiter," she says. We worked on the aerobraking process, which is how the orbiter achieves the right orbit when it reaches Mars. You keep dipping it in and out of the atmosphere to slow it down, so you're monitoring how far into the atmosphere it goes, how hot it gets, then you process that data and send it back to Mission Control."

Advice to Students

"It's a tough major," Schroeder says about the field she finally chose, "but you should stick with it. You should also find a network of some sort, whether it's other students or professionals. You'll need that support to get through."

Schroeder found support through the Society of Women Engineers, which she remains active in to this day as president of the Wichita chapter, participating in outreach programs for middle school students, among other things. "There weren't a lot of women in my classes," she says, "and it's sometimes hard to find other students to work with. I would have been totally lost without that group."

SUMMARY

- Milestones seem to be coming at exponential rates. For centuries, fire and rudimentary simple machines were state of the art technology. Over the span of a couple hundred of years, the steam engine transformed industry and transportation.

- The time span from the first flight of the Wright brothers to that of the Space Shuttle was not even 90 years. Technological innovation will continue to allow us to develop aircraft and spacecraft capable of performing feats previously considered impossible.

- In the modern classroom, most students carry a calculator with more computing power than the *Apollo* spacecraft took to the Moon. In the span of a decade, the U.S. space program progressed from barely putting a man in space to several missions on the Moon—a feat so monumental that a few people continue to claim it's a hoax.

- Often science fiction is the precursor to science fact. Much of the make-believe technology of the 1960s and 1970s sci-fi shows, such as personal communicators, has become today's reality. Where space travel brought dreams of flying cars and Stanley Kubrick's space station of *2001: A Space Odyssey,* of exploration and colonization of space, our present technology creates possibilities that stir the imagination. Humans are driven to explore the unknown and to travel to wherever nature allows.

- The technology and the aircraft developed along the way—miniaturization of electronics, predictable flight characteristics, sophisticated propulsion systems, autonomous guidance systems, even Velcro—have become necessities of our daily life. Tsiolkovsky may have been correct in thinking humanity is like a bird in a nest. We are capable of overcoming great challenges when bright minds work together. If we are to truly unlock our creative potential and prevent stagnation as a species, we must leave the nest and further explore our universe.

BRING IT HOME

1. Briefly describe the major innovation(s) developed by each of the following inventors:
   - Aristotle
   - Montgolfier brothers
   - Robert brothers
   - Daniel Bernoulli
   - George Cayley
   - Otto Lilienthal
   - Octave Chanute
   - Orville and Wilbur Wright
   - Samuel Langley
   - Frank Whittle and Hans von Ohain
   - Konstantin Tsiolkovsky
   - Robert Goddard
   - Wernher von Braun

2. Create a poster board that summarizes one aerospace innovation. Describe the inventor, the innovation, and how it improved the airplanes or spacecraft that followed.

EXTRA MILE

1. Construct a model kit of an airplane or spacecraft that interests you.
2. Construct a flying model airplane or rocket. (Safety guidelines for actual flight can be found at the following websites. You should always follow safety procedures when operating model aircraft and rockets!)
   - National Association of Rocketry (www.nar .org/)
   - Academy of Model Aeronautics (www .modelaircraft.org)
3. Build a full-scale mockup of a *Mercury*, *Gemini*, or *Apollo* space capsule out of cardboard.
4. Design a history of aviation mini-museum in a display case.

CHAPTER 3
Basic Aerodynamics

START LOCATION	DISTANCE	END LOCATION

Menu

Before You Begin

Think about these questions as you consider the concepts in this chapter:

1. How is the motion of an object affected by a force?

2. What is meant by the term "flying"?

3. How is lift generated?

4. What other forces affect the flight of an aircraft?

5. How is laminar flow different from turbulent flow?

6. Why are wings shaped the way that they are?

7. What is the difference between an airfoil and a wing?

8. What major components make up a typical airplane?

9. How does a wind tunnel work?

10. How has computer technology changed the way that aircraft are designed and tested?

Understanding why an airplane flies and how it is maneuvered through the air starts with an understanding of the four forces of flight. Throughout a flight, the four forces on an airplane are lift, weight, thrust and drag (see Figure 3-1). Knowing how to create and control the four forces is at the heart of aerospace engineering.

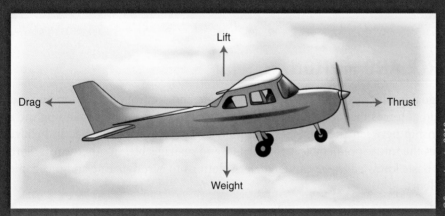

Figure 3-1: *Four forces on an airplane.*

Lift is the force created by the flow of air around the wing and supports the aircraft in the air. Changes in lift forces allow the pilot-in-command to maneuver the airplane through the air. **Weight** is caused by gravity and always acts downward toward Earth. **Thrust** is the force created by the aircraft's powerplant that propels it into the wind. **Drag** is the resistance force the aircraft experiences due to its passage through the air.

The task of an aerospace engineer is to create a well-designed aircraft that meets or exceeds the design constraints for how many passengers and crew will be on board, how much cargo needs to be lifted, how fast and far the craft will fly,

Lift:
in level flight, the upward force created by buoyancy and aerodynamic interaction with the air surrounding the aircraft. Lift forces can be varied and controlled to stabilize and maneuver the aircraft.

Thrust:
the force produced to overcome drag and momentum to control the velocity and position of the aircraft.

as well as how maneuverable or stable the aircraft needs to be to be useful and safe.

The focus of this chapter is the basic principles underlying the creation of the aerodynamic forces of lift and drag and how an airplane is constructed to produce these forces. Future chapters will take a closer look at stability and maneuverability and the function of the aircraft's powerplant.

Courtesy of NASA/Dryden Flight Research Center

NASA's F-15 Advanced Controls Technology for Integrated Vehicles (ACTIVE) awaiting testing in 1995. This vehicle used two nozzles that could turn up to 20 degrees in any direction. The new nozzles gave the aircraft thrust control in the pitch (up and down) and yaw (left and right) directions, reducing drag and increasing the aircraft's fuel economy and range.

FORCE AND MOTION

The goal in any mechanical system is to create and control the location and magnitude of forces that affect the future motion of the object. Understanding the relationship between these pushes and pulls and the resulting change in behavior makes the future predictable, controllable, and manageable. In other words, with a little basic knowledge, we can engineer solutions to aerodynamic and mechanical problems.

Newton's Laws

Sir Isaac Newton was the first person to publish a set of rules that accurately described the relationship between forces and motion. He summarized this understanding in three fundamental laws of physics that we refer to today as Newton's three laws of motion. The three laws apply just as well to airplanes, helicopters, and spacecraft as they do to any other object in motion.

Newton's First Law of Motion: Inertia

*An object in motion tends to stay in motion unless acted upon by a net force: **inertia**.*

Inertia:

an outcome of mass; all objects tend to continue moving in their current direction of travel unless acted upon by an outside force.

We can pull out many details from this simple statement. The first is that there are two motions that people immediately recognize as being distinct, moving and not moving (physics teachers refer to this as being at rest). Newton's first law doesn't distinguish between the two because there is nothing special about the state of rest. So for us, rest is just another motion like any other.

Accepting that there really is only one characteristic called motion allows us to shift our focus to the second detail in this law. Objects continue to do what they are already doing. The "natural" motion in the future is whatever you are already doing. So if you are just sitting there, that is your future.

The third detail of Newton's first law is that it describes a way to control the future motion of the object. The key words are "net" and "force." Forces are easy to understand; push on something, and it moves, right? Do a few quick thought experiments. What would happen if you push really hard on the walls of your room? Would they move or continue to just sit there? Roll a ball across a really smooth floor. When you stop pushing on it, does it keep rolling? The counter-intuitive results are due to an incomplete identification of all of the forces that are pushing on the object. At a minimum, the world is filled with frictional forces that exist whenever one substance tries to move past another substance. Thus friction, or drag, can bring an object to rest or keep it at rest.

Finding the Net Force

Summation of Forces with Trigonometric and Graphical Methods

We can find the net force by summing all of the individual forces together. This can be accomplished with trigonometry or scale diagrams (see Figure 3-2). To solve the problem with trig, each **vector** is broken down into its x, y, and z components, and then each is summed separately. The final result can be reassembled with trigonometry to find the net force. For example, find the net force equivalent for two forces where force #1 is a 10 lb force exerted in the 45° NE direction and force #2 is a 10 lb force exerted due east in the 90° direction.

Vector:

an arrow that represents both the direction and magnitude of a value by its orientation in space and the length of the arrow.

Figure 3-2 *Vector addition using trigonometry and graphical techniques.*

Trigonometric Solution:

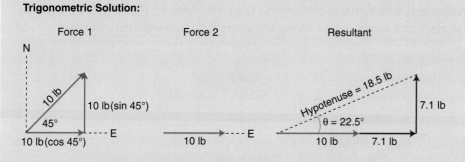

Graphical Solution:

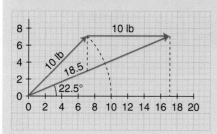

© Cengage Learning 2012

Find the northerly component of force #1:   10 *lb* (sin 45°) = 7.1 *lb*

Find the easterly component of force #1:   10 *lb* (cos 45°) = 7.1 *lb*

Sum the easterly components of both forces: 7.1 *lb* + 10 *lb* = 17.1 *lb*

We can find the resultants angle:

$$\theta \; from \; east = \tan^{-1}\frac{opp}{adj} = \tan^{-1}\frac{7.1 \; lb}{17.1 \; lb} = 22.5°.$$

Find the resultants magnitude: $Hyp = \dfrac{opp}{\sin 22.5} = \dfrac{7.1 \; lb}{0.383} = 18.5 \; lb.$

The solution can be found graphically by drawing the first force starting at the origin and drawing each subsequent force starting at the tip of the arrow for the preceding force. Then directly measure the magnitude and direction of the resultant from the origin to the final arrowhead.

The concept of a net force allows us to take all of the forces on an object and add them together so that they act as one force. Acting as if an airplane produces one net lift force makes it easy to calculate the total weight that it can lift (see Figure 3-3). However, this approach isn't very useful for designing the main beam, called a spar, that runs through the wing. A professional engineer designing the spar needs to keep all of the forces separated to design the part to be strong while minimizing the weight of the part.

In summary, according to Newton's first law of motion, if an object keeps doing what it was doing, then we know that either there are no forces on the object, or there are multiple forces that when added together completely cancel each other out so that the net result is the same as no force at all. If an object is changing its motion, then we know that there is an unbalanced force on the object.

Figure 3-3 The lift force per foot of wingspan is shown for an aircraft entering a left turn. Note that the lift per foot of span is less on the aircraft's left wing (right in diagram), which causes the roll into the turn. However, the total lift is equal to the aircraft's weight, which allows the aircraft to turn while maintaining altitude.

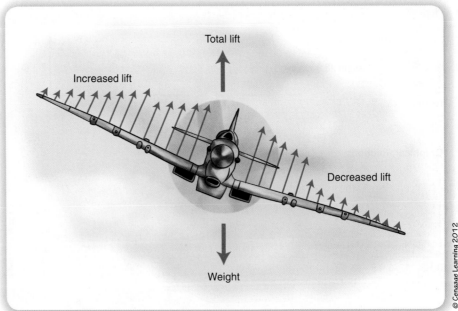

© Cengage Learning 2012

Newton's Second Law of Motion: Acceleration

An object will experience an acceleration in the direction of the net force that is proportional to the net force on the object and inversely proportional to its mass.

As soon as we observe that an object is changing its motion by speeding up, slowing down, or moving along a curved path, we know that it has a net force applied to it. The change in speed or direction is called acceleration. How rapidly the speed of an object changes, or how tight of a curved path the object is following, is determined by both the size of the net force applied and the mass of the object (see Figure 3-4).

The second law matches well with our experiential knowledge. Causing a change in speed, or acceleration, due to force makes sense. The larger the change in speed you want to produce, the more forcefully you have to exert yourself. If you want to stop your bike more quickly, you need to squeeze harder on the brakes.

Mass is the amount of material that an object contains. On Earth, we frequently measure the force of gravity on an object, which is called weight, to determine the mass of the object. The electronic balance, or scale, is calibrated to divide out the gravity effect and leave only mass.

Consider the following example. A small child is sitting in a driveway when a vehicle starts to roll down a slope toward them. We will assume that how you react depends on the vehicle that is doing the rolling. For a little red wagon, you might jump in and grab the wagon, but for a massive truck, you would grab the child. Why the difference in your reaction (see Figure 3-5)?

In both cases, the force you can exert is based on the strength of your muscles, so you should be able to produce the same acceleration based on force alone. However, the relationship to mass is another story. The wagon is very low mass, so your force produces a rapid change in speed. It's a safe call to assume you can stop the wagon before anyone is injured. However, the truck is 1,000 times more massive, so it will experience a 1,000 times slower rate of change in its speed. (We will assume that the vehicle is no longer on a slope or other factors would matter as well.) So it's a safer option to grab the low-mass child and assume that you can speed them up in much less time than it would take to slow down the truck.

In summary, how quickly an object can change its motion depends on both how hard it is being pushed and how much mass the object contains.

> **Acceleration:**
> the rate of change in velocity.

Figure 3-4 Friction with an atmosphere provides aerobraking, which slows a satellite.

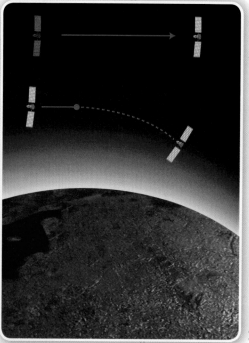

© Cengage Learning 2012

Figure 3-5 Mass and inertia.

© Cengage Learning 2012

Newton's Third Law of Motion: Action and Reaction

For every action, there is an equal, but opposite, reaction.

Expanded slightly, Newton's third law states that for every force acting on an object, there is a force of equal magnitude exerted on the source of the first force but in the opposite direction. This means that forces come in pairs. Modify our first law thought experiment to push against a large object such as your bed. If you push on the bed, it pushes back into you; you can feel the pressure on your hands and muscles. Push harder and the bed pushes back harder, that is the requirement to meet the equal but opposite part of Newton's third law. But if the forces are equal, how can you ever move the bed? The key is that the reaction force is created on another object. Up to a point, you can push harder and harder but nothing happens as both you and the bed also have a **friction force** with the floor of the room. However, as soon as one of the paired forces is greater than the friction with the floor (for that object), then it starts to move (see Figure 3-6)! The forces are equal and opposite, but they don't cancel out as they are applied to different objects. In many engineering situations, we design a way to create the action force so that the reaction causes the change that we want accomplished. Collectively, Newton's laws of motion allow us to predict and control the motions that will be produced by the forces applied to our aerospace designs.

Figure 3-6 Action and reaction forces, mass, and inertia.

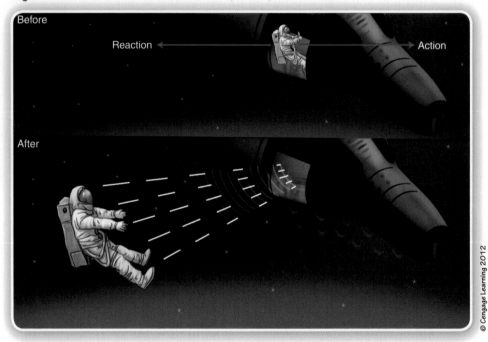

Motion: Linear Versus Rotation

Two distinct types of motion can be changed by a force. **Linear motion** is along a straight line, whereas **rotation** occurs around a point in space called the object's center of gravity. Understanding the relationship between the center of gravity and a force determines how the motion of an object changes over time.

Before we can really understand linear and rotational motion, we need to understand the meaning of the term **center of gravity**. In simplest terms, the center of gravity (c.g.) is the point at which we can act as if the entire weight of an object exists. We can find the c.g.'s location mathematically for simple shapes because it's

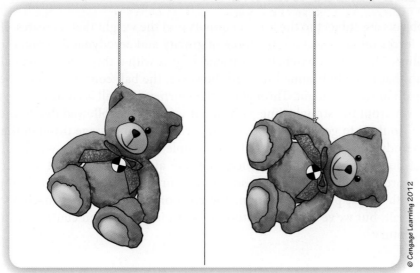

at the center point of symmetrical objects. Experimentally, the c.g. will always come to rest directly under any point from which we suspend the object (see Figure 3-7).

Linear Motion, Velocity and Acceleration

After the c.g. has been determined, it defines the relationship between a force and its resulting acceleration. If the net force points directly through the c.g. location, then all of the force goes into changing the velocity of the entire object (its speed and/or direction) without inducing any rotation. The object will shift its location in space while maintaining its orientation or rate of rotation in space.

Rotational Motion, Velocity, and Acceleration

If a net force is applied so that it does not point directly through the c.g., then the force will result in a **torque**, or twisting force, which will cause the object to change both its location and orientation in space. The same force, applied in different locations, can produce dramatically different changes in the motion of an object (see Figure 3-8). This distinction will be very important in later chapters when we consider the operation of an aircraft as a whole.

Figure 3-8 Force applied through the CG leads to linear motion; anywhere else leads to some rotation.

FLYING VERSUS FALLING

All objects are subject to the force of gravity and the weight that it creates. If the only forces on an object are the force of gravity and aerodynamic drag, then it is falling. Throwing a baseball to a friend begins with a throw of the ball along a path that initially is aimed upward; however, the ball continually falls below this path until it is caught. Throughout the journey, the ball is continuously falling away from the straight-line path that it would have followed due to inertia alone. To fly is to generate aerodynamic forces due to passage through the air. The aerodynamic force can partially or completely overcome the weight of the object due to gravity, which reduces the rate at which the object accelerates downward. This lifting force can sustain the aircraft in flight longer than possible with simple projectile motion. The creation of lift changes a machine into an aircraft, but we've only understood how lift is created and controlled in the past century.

METHODS FOR CREATING LIFT AND OTHER AERODYNAMIC FORCES

An airplane's movement through the air leads to the creation of lift forces in four primary ways: deflection, Bernoulli, Coanda, and circulation processes.

Deflection Forces

Hold a ping-pong paddle in the wind, and you know that even a flat board can generate an aerodynamic force. Deflection lift is produced indirectly when a surface moves relative to the air at an angle that is between parallel and perpendicular to its direction of motion. It doesn't matter whether it is the air or the board doing the moving, the only requirement is that air molecules strike the surface at an angle and bounce off in a new direction.

To produce an upward force, the board needs to be horizontal with the leading edge of the board higher than the trailing edge so that air molecules are deflected downward as they strike the bottom of the surface. Our observation that the air molecules have changed their motion downward implies that they encountered a net force (first law) and that the surface pushed on the air molecules in the downward direction (second law). This requires that the surface experiences an equal but opposite (third law) reaction force that is an upward acting lift (see Figure 3-9). We can create deflection forces on any surface that is struck by moving air. The deflection force is always in the opposite direction of that in which the air molecules have been deflected.

Bernoulli

Every explanation for how an airplane can fly discusses Bernoulli lift, and yet Daniel Bernoulli lived hundreds of years before an airplane achieved sustained, controlled, powered flight with a human on board. Bernoulli was a scientist and mathematician concerned with the fundamental laws of fluids in motion. His famous equation is simply an expression of the conservation law of energy, and yet it creates an opportunity to engineer an outcome by predicting in advance the amount of lift a wing will create due to the increased velocity of the airflow over the wing.

At its heart, the Bernoulli equation states that a fluid (like air or water) contains energy due to its static pressure, motion, and position. The sum of these

Figure 3-9 *Deflection of airflow by a flat ping-pong paddle.*

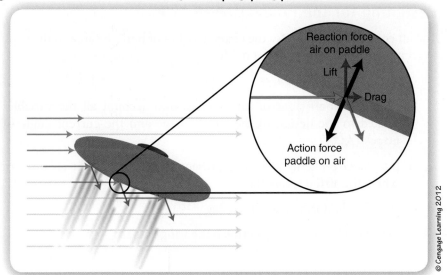

Reaction force
air on paddle

Lift

Drag

Action force
paddle on air

© Cengage Learning 2012

internal, **kinetic**, and **potential energies** is a constant in a closed system, which is any case in which energy isn't added or removed from the air. Because energy is conserved, if the energy in one of the three categories is increased, it has to result in a decrease in some other category. Because the difference in height from the bottom to the top of an airfoil is insignificant, the exchange is between the static pressure and the kinetic energy due to the velocity of the airflow. Bernoulli's equation states that if we increase the velocity of a fluid, the fluid will exert less pressure (force per unit area) on the surface over which it is flowing.

What does this have to do with flying? We need to understand one more piece before Bernoulli lift can be completely understood. At low speeds, air acts nearly incompressible, meaning that it doesn't change its density significantly due to being pushed around; it simply flows out of the way of a moving object. This means air obeys a law of continuity for flow that states that when it flows through a reduced area (such as a smaller pipe), it has to speed up to have the same volume of air pass through every location in the system (see Figure 3-10). An airfoil is an engineered shape designed to produce such an increased velocity and its reduction in pressure over the wing.

Bernoulli's equation provides a way to calculate the difference in pressure between the lower and upper surfaces of the wing. Multiplying the pressure difference by the area of the wing results in the total lift created due to the velocity differences produced by the airfoil. However, this is not the total lift created by the wing.

The full lift equation includes terms that define the effectiveness of the shape at producing lift (coefficient of lift, C_L), the impact of the density of the air (the Greek letter *rho*, ρ), the velocity at which the air is flowing across the airfoil shape (velocity, V), and the total area over which the lift is being generated (planform area, A). Although the actual lift generated on any given part of a wing is unique, the lift equation gives us a means to predict the total lift that a wing can generate.

Kinetic energy:

the ability to cause change due to the speed of an object. Energy is stored as kinetic energy when the speed of the object increases and is released when the object slows down.

Potential energy:

the ability to cause change due to the position of an object and its ability to fall at some time in the future.

Figure 3-10 *Airflow velocity and pressure in a venturi.*

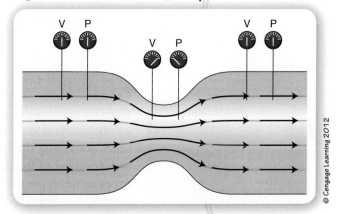

© Cengage Learning 2012

Designing for the Production of Lift

The lift equation incorporates the characteristics of both the air and the wing to accurately predict the amount of lift the wing can produce:

$$L = \frac{1}{2} \cdot C_L \cdot \rho \cdot A \cdot V^2$$

▶ C_L is the **coefficient** of lift. It takes into account all the variables related to the design of the airfoil shape and the current **angle of attack (AOA)**.

▶ ρ is the Greek letter *rho* and represents the density of the air through which the airfoil is moving.

▶ A is the planform area of the wing.

▶ V is the relative velocity between the wing and air.

Example

How much lift can a small four-seat aircraft produce? Follow this calculation for a typical aircraft.

Cessna 172R Takeoff at 75 KIAS (64 MPH) with 10° Angle of Attack		
Term	**English (Lift in Pounds)**	**Metric (Lift in Newtons)**
C_L	1.458	1.458
ρ	0.00241 slugs/ft³	1.242 kg/m³
A	174 ft²	16.2 m²
V	94 ft/sec	28.6 m/sec

$$Lift = \frac{1}{2} \times C_L \times \rho \times A \times V^2$$

$$Lift = 0.5 \times 1.458 \times 0.00241\frac{slugs}{ft^3} \times 174\,ft^2 \times \left(94\frac{ft}{sec}\right)^2$$

$$Lift = 2700\ lbs$$

Notice the units of measure for each term. Practice making the metric calculation; its solution is 12,000 Newtons.

Coanda

Place a curved surface such as a spoon into a thin stream of flowing water, and it's obvious how the inner surface of its bowl causes a deflection of the stream of water away from the spoon's surface and, by the third law of motion, a pushing of the spoon out of the stream of water. What is more challenging is to observe what happens when the outer convex curvature of the same spoon is brought near the stream. The spoon is drawn into the flow!

GRASS STRIP ADVENTURE

Watch this You Tube video of the Coanda spoon experiments:
www.youtube.com/watch?v=AvLwqRCbGKY

The Coanda effect is the result of the passage of the fluid over the curved surface of the spoon. As the air flows, it induces two different results in the air nearby. The first effect is the airstream has an increased velocity, which reduces the local pressure and draws in additional air to join the airstream. This Bernoulli effect enhances the results by involving more air in the process. The second effect is to create a rarefication (reduction in density) of the air very close to the surface of the spoon or airfoil as the airstream draws air molecules out of this region.

Just as energy conservation explains why the Bernoulli equation works, momentum conservation explains the Coanda effect. Momentum represents the overall motion that is contained in an object or system. Conservation simply means that if one part of the system increases its motion in one direction, then some other part of the system has to increase its motion in the opposite direction.

For an airfoil, the result of the Coanda effect is that the airflow above the wing is induced to follow the curvature of the wing's upper surface so that it leaves the trailing edge of the wing moving downward in what is referred to as downwash. The overall motion of the air downward due to the Coanda effect represents a momentum change that requires an equal upward reaction from the wing itself.

Circulation

An airfoil in motion through air induces an upwash in front of the leading edge of the wing and a downwash behind the trailing edge of the wing. If a camera is fixed relative to the airfoil the result is observation that the wing is stationary as the air flows past the wing and that the upwash and downwash locations remain fixed in position relative to the wing. However, if we fix the camera in place relative to the freestanding air, it will look like the wing is moving through the air and that a region of upwash one moment will briefly flow horizontally and then become a region of downwash. The observation can be summarized as a bound vortex that moves with the wing.

The bound vortex appears as a cylinder of air circulating upward at the leading edge and downward at the trailing edge of the wing. The resulting circular rotation represents an angular (or rotational) momentum change in air that previously was not moving. The conservation of momentum law extends to rotation as well, which means that the downward descending air at the trailing edge needs to have a circular motion that is opposite of that in the bound vortex. The end result is an enhanced airflow pattern around the wing with an induced increase in velocity over the wing and induced decrease in velocity below the wing. Vortex generation therefore indirectly reinforces the creation of Bernoulli lift by an airfoil shape.

Center of Pressure

The air surrounding a wing exerts pressure everywhere on its surface. The amount of pressure exerted varies from location to location. If the total pressure on the bottom of the wing exceeds the total pressure on the top of the wing, then the wing is producing lift. To simplify the situation, we can add all of the individual pressure vectors, everywhere, together so that we can act as if there is only one lift vector acting at the location known as the center of pressure. If the center of gravity and center of pressure are at the same location, the aircraft tends to maintain its motion because there is no net torque to cause the nose to pitch up or down. If the center of pressure is in front of the center of gravity, the aircraft tends to pitch nose up and vice versa pitch down when the center of pressure is behind the center of gravity. In real airplanes, we balance out any tendency to rotate around the center of gravity by shifting weight in the aircraft or creating other aerodynamic forces so that the pilot can decide how to maneuver the aircraft.

Vortex:
a rotating cylindrical mass of air. Vortexes tend to spread, descend, and dissipate over time.

For wings that use a **symmetrical** airfoil, the center of pressure can be found at about 25% of the distance from the leading to the trailing edge, or the quarter-chord point, of the airfoil's shape. This location doesn't move as the airfoil is maneuvered to different AOAs (coefficients of lift), so the aircraft tends to continue the maneuver after it is started. This is very desirable for aerobatic aircraft.

The center of pressure for **asymmetrical** or cambered airfoils like those found on most general and commercial aviation aircraft tends to start much farther back when at low AOAs and moves forward as the airfoil's AOA is increased. This has a destabilizing effect that tends to increase the rate of a maneuver in pitch. This tendency is overcome by adding a compensating negative AOA to the horizontal stabilizer to improve stability.

So what about flying wings? To avoid the issues of a conventional cambered airfoil's need for other surfaces, flying wings use a reflex-cambered airfoil shape in which the camber line curves upward near the trailing edge (see Figure 3-11). This change makes the trailing edge portion of the wing act as if it were the stabilizer for a conventional cambered airfoil.

Figure 3-11 *The stabilizer's required incident angle of attack (AOA) depends on the shape of the main airfoil.*

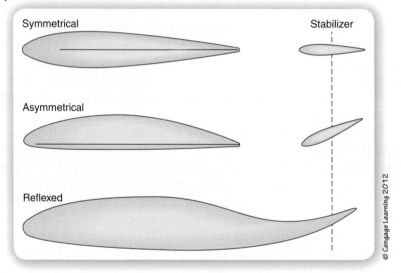

© Cengage Learning 2012

Lift and the Angle of Attack (AOA)

The creation of lift is a product of an interaction between the physical structure of the aircraft and the flow of the air around the structure. This interaction can be summarized by a single term, the coefficient of lift. This one term summarizes all of the complexity of the interaction and is typically directly measured from wind tunnel experiments.

The Lift Coefficient (C_L)

The **coefficient** of lift captures the effect of both the shape of an airfoil and the AOA at which it is presented to the air. For a given airfoil shape and AOA, there will be one coefficient of lift. As we've already seen, this plays a major role in the production of lift because it's a term in the lift equation. The coefficient of lift tends to increase linearly (in a straight line on a graph) as the AOA is increased (see

Figure 3-12). This implies that one way to produce more lift with the same airfoil is to increase its AOA. Pilots constantly adjust the AOA of wings and controls to change the amount of aerodynamic forces that are produced so that they can maneuver and control the aircraft.

Stalls

The coefficient of lift increases linearly with AOA only up to a point. For most airfoil shapes, the critical AOA at which the air can't flow uniformly (**laminar flow**) over the airfoil shape and begins to tumble (**turbulent flow**) is about 18°. Beyond this point, the coefficient of lift rapidly decreases to zero, and the wing is described as being in a stalled condition (see Figure 3-13). The critical AOA can vary greatly from one airfoil shape to another.

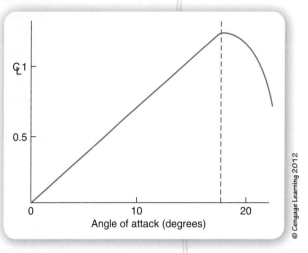

Figure 3-12 A typical *coefficient of lift* graph for an airfoil.

© Cengage Learning 2012

Laminar flow:

the flow of a fluid in parallel layers in which each layer, or streamline, has a unique velocity, but there can be relative motion between layers.

Figure 3-13 Angle of attack (AOA) and laminar versus turbulent flow.

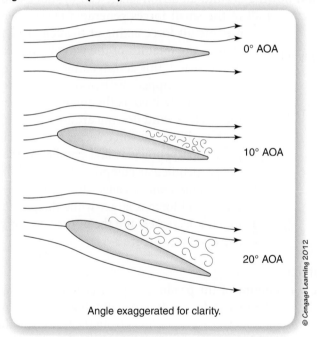

0° AOA

10° AOA

20° AOA

Angle exaggerated for clarity.

© Cengage Learning 2012

This concept has an important outcome for pilots. If a wing is stalled, it is producing very little lift, and the aircraft acts as a projectile. Given enough time, slow-moving projectiles in the atmosphere strike the earth. For an airplane, that is generally not a good thing. To be flying, the wing needs to produce lift. To maintain control, a pilot has to keep the AOA below the critical angle.

Stated another way, an airplane can exceed the critical AOA while in any attitude (orientation in space) and at any airspeed; the only thing that matters is the AOA. Pilot training includes many experiences that demonstrate what it feels like when the aircraft is about to stall and proper recovery techniques so that the pilot can safely avoid stalls and fly out of the stall quickly should one occur.

CREATING DRAG

If lift is required to overcome the weight of an aircraft so that it can fly, then drag is the force that tries to bring the aircraft to rest. Some situations call for very large drag forces, such as when a large airplane needs to slow down to land on a short runway, and other situations need very little drag, such as when soaring in a glider. To design an aircraft to produce the amount of drag desired, we need to understand how drag is produced.

Induced Drag

The creation of lift automatically induces the creation of drag. The more lift that is produced, the more drag that is produced. So whether it is due to a change in velocity or AOA, when a pilot maneuvers quickly (which requires large lift forces), the aircraft also tends to slow down quickly. Fighter pilots talk about energy management and train extensively to learn how to maneuver their aircraft while maintaining the speed necessary to achieve their mission.

The Drag Component of Lift

A relative wind flowing around an airfoil shape can produce a force. We know that we can act as if all of the small forces generated everywhere on the wing are one lift force acting at the center of pressure. If we change the total lift of the wing by increasing the AOA of the airfoil, this aerodynamic vector grows in magnitude and is tipped farther and farther rearward. However, lift is defined as the force that is perpendicular to the relative wind flowing around the wing.

Figure 3-14 **Lift and drag versus total aerodynamic force.**

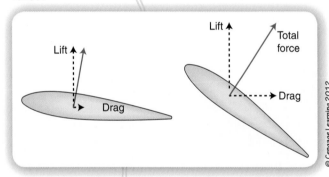

© Cengage Learning 2012

If we break the force vector for the airfoil into a component perpendicular and parallel to the relative wind, the perpendicular component is called lift, and the parallel component is called drag. Increase the AOA, and the overall force is larger while the portion that is going into lift decreases, and the portion that is going into drag increases (see Figure 3-14).

Wingtip Vortices

If we change the amount of lift produced by an airfoil by changing our velocity through the air, then we induce a drag penalty due to the creation of wingtip vortices. We already know that as the air flows from the leading to the trailing edge, it induces Bernoulli lift by reducing the pressure above the wing. The resulting normal (or elevated) pressure below the wing pushes upward on the wing while the reduced pressure above the wing pushes downward less strongly. The result of this push-of-war is Bernoulli lift.

However, out near the wingtip, the air has another option. In addition to flowing from leading to trailing edge, the high pressure air below the wing can take a shortcut and flow upward around the wingtip. The conservation of momentum law requires that this upward flow at each wingtip be compensated for with some air somewhere else moving downward. Combine this with the forward motion of the aircraft through the air, and the result is the formation of very large, expanding spirals of air trailing off of both wingtips (see Figure 3-15). These wingtip vortices rob energy from the system and thus act as drag that brings the aircraft to rest.

Reducing the formation of wingtip vortices can be achieved with winglets, tapering of the wing, increased aspect ratio, and combinations of techniques.

Figure 3-15 Wingtip vortices from a C-17 Globemaster III flight through smoke from a flare.

Courtesy of U.S. Air Force

Drag reduction can be remarkable. The next time you are at the airport, look and see how many airplanes use one or more of these methods.

Parasitic Drag

Parasitic drag is the result of the airplane passing through the air and rubbing against individual air molecules creating a frictional force. Airplanes that regularly operate off of grass strips and unimproved airfields can have a significant amount of bugs and dirt adhered to their surface, which increases the roughness of the surface. Frictional drag is directly related to the total amount of surface area (or wetted surface) of the airplane and the smoothness of that surface. A much larger mass of air has to be shoved out of the way of the oncoming aircraft's profile or shape because two substances can't occupy the same place in space at the same time. This profile drag is directly related to the total cross-sectional area (the area of a slice across the aircraft) of the aircraft as well as how that area changes from the nose to the tail of the aircraft. At high speeds, the aircraft generates another source of drag because the air in front of the aircraft does not have time to flow out of the way of the oncoming aircraft without changing its density by compressing closer to its neighboring molecules. In sum, numerous factors not related to the production of lift can induce the creation of parasitic drag forces.

Ground Effect (Ekranoplans)

When an airplane flies very close to the ground, typically within a half wingspan of the surface, there is a marked reduction in the amount of upwash, downwash, and wingtip vortex generation. This means that the aircraft produces much less drag while it continues to produce the same amount of total aerodynamic force. If the total is the same, but less of it is drag, then more of it must be lift.

The impact of **ground effect** can be significant as an overloaded aircraft may be able to achieve flight in ground effect and then be unable to climb above obstacles at the end of the runway. During approach for landing, an airplane with too much speed may float in ground effect and be unable to stop on the remaining runway after touchdown.

Ground effect:

the increased lift and reduced drag experienced by an aircraft when it is operating within one wingspan of the ground (or surface).

Case Study ⫸→

RAPID TRANSPORTATION ACROSS INLAND SEAS

In the former Soviet Union, numerous aircraft have been specifically designed to "fly" in ground effect to rapidly deliver massive cargos across the Caspian Sea. These specialty aircraft, known as ekranoplans, never fly more than a few hundred feet above the tops of the waves yet are capable of very rapidly delivering large amounts of cargo very efficiently (see Figure 3-16). The ekranoplan offers the advantages of an airplane's rapid transport with the ship's fuel efficiency.

Ekranoplans, or ground effect aircraft, are only effective above very flat land or water as any significant variation in height tremendously increases the chances for a collision with the surface.

Two ekranoplans could be seen on Google Earth located at 42°52′54″N, 47°39′24″E and at 42°52′50″N, 47°39′57″E. A structure on a nearby beach may be a third disassembled ekranoplan.

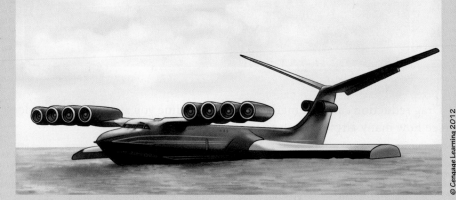

Figure 3-16 *An ekranoplan aircraft designed specifically for flight within ground effect.*

© Cengage Learning 2012

LAMINAR AND TURBULENT FLOW

Laminar, or streamline, flow describes the movement of a fluid along parallel paths. This means that the fluid is not disturbed or mixed due to the flow. Turbulent flow is characterized by a chaotic mixing of the fluid due to the flow. In general, laminar flow is associated with less drag and more lift, whereas turbulent flow produces a lot of drag and very little lift.

Boundary Layer

Because air is viscous, it tends to stick to a surface. This results in the creation of a thin layer in which the velocity of the air flow transitions from no relative motion at the skin of the aircraft surface to the freestream velocity some distance away from the surface. Boundary layers are important because flow starts laminar at the leading edge and becomes turbulent at some point on the airfoil. This is due to surface imperfections, curvature of the airfoil shape, pressure gradients, and other factors. The end result is that by controlling when flow transitions from laminar to turbulent flow, we can design for reduced drag and greater lift, two of the primary goals of the aerospace engineer.

Stalls

We've already discussed how a stall occurs whenever a wing is moving at or beyond its critical AOA. This is related to a discussion of laminar and turbulent flow due

Stall:

beyond the critical AOA, an airfoil rapidly reduces the amount of lift that it produces. This is a stalled condition.

to the expanding zone of turbulent flow at the trailing edge of the wing as AOA is increased. Up to a point, the air is capable of following the contour of the upper surface of the wing and maintains laminar flow and lift creation capability. However, as the critical AOA is reached, the place on the upper surface at which the turbulent flow begins to occur moves quickly toward the leading edge. When the turbulent flow approaches the region from 25 to 50% of the chord, where most of a wing's lift is produced, the production of lift collapses, and the wing is stalled. Aircraft designers use numerous tricks to control flow in the boundary layer and to ensure that the wing has predictable and safe stall characteristics.

AIR FOIL SECTIONS

An airfoil is a shape designed to transform motion through a fluid into a force.

Terminology

To understand airfoil design, we need to have a common vocabulary that describes the features of the shape. This terminology relates to the two primary characteristics of an airfoil: the curvature of its surfaces and the locations of features within the shape.

The most important locations on an airfoil are the leading and trailing edges that describe the first and last place at which the airfoil directly contacts the airflow around the shape. They are also where stagnation points occur where there is effectively no airflow velocity relative to the surface of the airfoil as the flow splits or rejoins. The line that connects the leading edge to the trailing edge is referred to as the chord line (see Figure 3-17).

Figure 3-17 Airfoil terminoloy.

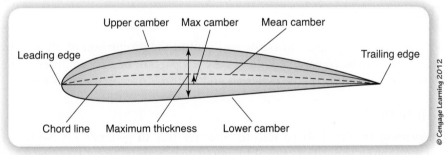

© Cengage Learning 2012

The length of the chord line is a critical dimension that is used to determine the full-size wing and as the parameter against which other dimensions in the airfoil shape are determined. In engineering software such as Autodesk Inventor, this concept is the foundation of parametric dimensioning. The angle from the chord line to the direction of the relative wind's flow is the AOA at which the airfoil is functioning.

Camber describes the curvature of a surface. For an airfoil, we are primarily interested in the amount of curvature in the upper camber and lower camber of the shape, which are expressed as a percentage of the total length of the chord line. Cutting perpendicular across the chord line, we can plot in locations that are equidistant from both surfaces of the airfoil. Connecting these points gives us the mean camber line, which is an effective predictor for the performance of the entire airfoil. The location of maximum thickness, maximum camber, and the curvature of the leading edge play major roles in determining the lift and drag that a particular airfoil will produce.

Symmetrical and Asymmetrical

A symmetrical airfoil has the same camber on the upper and lower surface, which makes the mean camber line a straight line that lies along the chord line of the shape. Symmetrical airfoils are useful when the aircraft is frequently operated inverted as the wing performs equally well right side up as it does upside down. Symmetrical airfoils must be flown at an AOA to produce any net lift force.

Asymmetrical airfoils typically have more curvature on the upper surface, which pulls the mean camber line into a curve that lies above the chord line. Asymmetrical airfoils produce lift even at an AOA of zero; however, they are much more effective at overcoming gravity in normal flight than they are when inverted. Almost all airfoils are asymmetrical with the exception of aerobatic and military fighter aircraft, which may use a symmetrical airfoil instead.

NACA Testing

The primary way to determine the effectiveness of an airfoil shape is to gather data directly while the shape is producing lift and drag. This is most easily accomplished in the controlled setting of a wind tunnel in which all of the other factors in the Bernoulli equation can be controlled and maintained throughout the experiments.

Starting in the 1930s, the National Advisory Committee for Aeronautics (NACA) undertook an intense research effort that published the first real performance data on a set of 78 related airfoil shapes. The publication of this data provided engineers and manufacturers with the starting place for their aircraft designs.

Throughout the next decades, NACA published the performance data and included coefficients of lift and drag at numerous AOAs for many hundreds of airfoil shapes. This data collectively formed the backbone of aerospace innovation for many decades.

WING CONSTRUCTION

A wing is a device designed to create and control the production of lift and drag forces. Wings can be constructed out of very few parts or be very complex in their construction depending on the materials and methods used in their design.

Components

Transforming an airfoil shape into an effective way to generate lift requires its integration into a wing. The critical functions that have to be achieved by a wing include producing lift, transferring lift to the rest of the aircraft, bearing the forces of flight operations, and providing storage and mounting points for other components of the aircraft (see Figure 3-18).

Producing lift requires that the airfoil shape is maintained and that the air is forced to flow around the shape smoothly. In most aircraft, ribs are used to define the shape. A **rib** is a physical assembly that braces the surface of the wing against the inward pressure of the air. The profile of a rib is that of the airfoil shape and frequently includes features to allow the **skin** of the aircraft to be attached directly to the rib. Ribs are attached to each other and braced against rotation and shifting by **spars** and the skin of the wing.

A spar is a beam that extends from the root of the wing toward the wingtip. The spar stiffens the wing against bending and transfers the lift force back to the fuselage. The spar has to provide upward support of the wing when it isn't flying and downward bracing to absorb the lifting force produced by the ribs and skin during flight. The spar can be stiff enough to fully support all of these loads by itself

Figure 3-18 *The internal structure of a Skybolt biplane.*

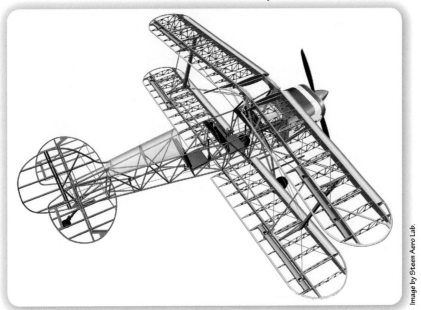

Image by Steen Aero Lab.

(internally braced wings) or may require the use of smaller components to brace them in place. Spars can extend into, or through, the fuselage to increase their ability to transfer the lift force to the fuselage and brace the wing panel in place.

Typical struts attach to the midpoint of the spars in each wing panel and the bottom of the fuselage frame. This forms a triangular structure that places the strut under tension when the aircraft is in flight. Wings can also be braced with wire, wooden poles, or hollow aluminum tubes. When on the ground, the weight of the wing puts the strut under compression. The strut itself may be braced with additional components to prevent the strut from bending under this force.

Frequently, the skin of the wing forms the leading and trailing edges of the aircraft as it flows from the top to the bottom of the wing panel. Leading edges may incorporate landing lights and lift-enhancing devices. Trailing edges incorporate movable control surfaces and lift-enhancing devices. Internally and externally, the wing can include mounting spaces for fuel tanks, engines, and landing gear.

Wing Planform

If you lay on the ground watching airplanes fly overhead, you will quickly recognize that the wings of the airplane have a distinct profile to their outline. This shape is referred to as the **planform** of the wing. Planforms vary wildly but are the result of a small set of characteristics that include span, chord, taper, sweep, aspect ratio, and elliptical shape:

▶ **Span.** The distance from one wingtip to the other.

▶ **Chord.** The distance from leading edge to trailing edge in the direction of flight.

▶ **Taper.** A change in chord from root to tip.

▶ **Sweep.** An angle of the leading edge or trailing edge to the perpendicular to the direction of flight.

▶ **Aspect ratio.** The square of the span divided by the wing area, or for constant chord wings, simply the span divided by the chord.

▶ **Elliptical shape.** A curved leading edge or trailing edge.

> **GRASS STRIP ADVENTURE**
>
> Many extreme duration aircraft have unusually long wings. You can check out a flight video on You Tube of takeoffs for the Voyager around the world flights and Helios at *www.youtube.com/watch?v= Ml73lXW_jLs*

Figure 3-19 Wing planform and stall characteristis.

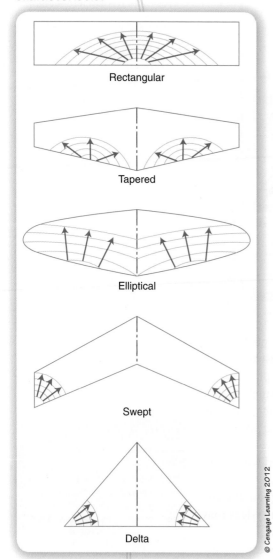

Rectangular

Tapered

Elliptical

Swept

Delta

© Cengage Learning 2012

Figure 3-20 Aspect ratio of a wing.

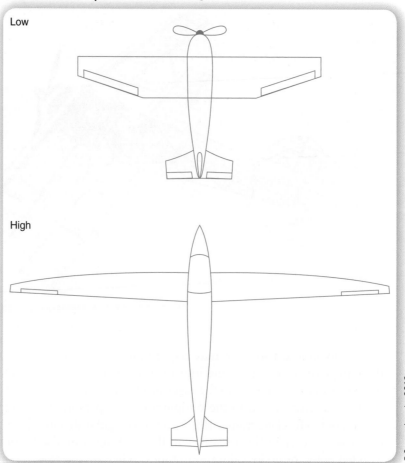

Low

High

© Cengage Learning 2012

Every planform has advantages and disadvantages (see Figure 3-19). Rectangular wings are cheap and easy to construct and have very gentle stall characteristics. Elliptical wings are difficult to manufacture and are expensive but greatly reduce drag. Swept wings are great for high-speed flight but can have dangerous tip-first stall characteristics. Like many things in aerospace engineering, designing for performance is a series of tradeoffs between competing benefits.

Aspect Ratio

Note that Bernoulli's equation does not have a term for aspect ratio. The production of lift is related to the shape of the airfoil and the total area of the wing's planform but has no direct connection to the wing's **aspect ratio**. On the other hand, the aspect ratio of an aircraft's wings is very strongly related to the efficiency with which it produces lift without inducing drag (see Figure 3-20). Gliders, which are some of the most efficient aircraft, tend to have very high aspect ratio wings. How aspect ratio leads to this high lift to drag ratio (L/D) is related to two factors: the amount of wingtip chord and how aggressively the airfoil shape produces a pressure difference.

To simplify the discussion, we will consider only rectangular planform wings and assume that the airfoil shape (not its size!) and the total area of the wings are constants. Thus, the total amount of lift produced should remain constant according to the Bernoulli equation. To understand the influence of aspect ratio on the L/D ratio, we will take it to the extreme as we consider its impact on the production of lift and drag.

Aspect ratio:

the ratio between the average chord of the wing and its span from wingtip to wingtip.

Imagine the standard Hershey bar wing of a typical training aircraft with a span that is approximately 8 times longer than the chord of the wing. This aspect ratio and airfoil shape will produce vortices of a given size at any particular velocity and AOA. Now increase the aspect ratio. For the area to remain a constant, the chord must decrease as the span increases. Continue this until the chord is very, very small, and the span is enormous.

These changes would have two major impacts. First, there is almost no wingtip, which means that the vortices they generate would be very small compared to the total lift produced thus producing very little induced drag. Second, the airfoil shape would be very, very small as well, which means that the air flowing around the airfoil would be experiencing very little change due to the passage of the wing, thus very little induced drag would result from the production of lift. In general, and up to a point, increasing the aspect ratio increases the efficiency of a wing.

Point of Interest

Why Aren't All Airplanes Designed with Long Wings?

High aspect ratio wings are efficient, so why aren't all wings designed with very high aspect ratios? Although efficient, long, thin wings have other limitations. To brace a long wing so that it doesn't drag on the ground or flex upward by too much requires much more strength and rigidity to handle the forces. Although the wing produces lower induced drag, the rate at which parasitic drag increases is even more rapid. In many aircraft, space inside of the wing is essential for fuel, and as the thickness of the wing decreases, this becomes impractical. Long wingspan aircraft are also quite cumbersome because they require larger hangar space, runways, and taxiways to operate. Finally, long wings give an aircraft a large inertial resistance to rotation. Any aircraft that needs to be nimble in roll requires shorter wings of lower aspect ratio.

Washout

Wings can be constructed so that the wingtip is at a slightly lower AOA than the root of the wing. This is accomplished by raising the trailing edge of the tip ribs slightly compared to those at the root. In flight, this ensures that at high AOAs, such as during slow flight and landing, the root portion of the wing panel will stall first with the stall spreading outward toward the tip as the AOA increases. The result is to reduce the tendency to tip stall, which induces a rapid snap roll and reduces the effectiveness of the ailerons. Some washout thus makes an aircraft much safer in the hands of less experienced pilots by giving them ample warning of an impending stall and allowing them to maintain control of the aircraft as a stall progresses.

Other Components

A complete wing can include numerous movable control surfaces in the form of ailerons, flaps, slats, or spoilers. The wing design and construction has to have locations for mounting these movable surfaces and for all of the linkages and mechanisms required to control them from the cockpit of the airplane.

Powerplant

The powerplant includes both the engine that transforms energy from one type to another as well as the device that converts this energy into a thrust force (see Figure 3-21). Thrust is required to overcome friction and drag to increase or maintain speed through the air.

Figure 3-21 **Major components of an airplane.**

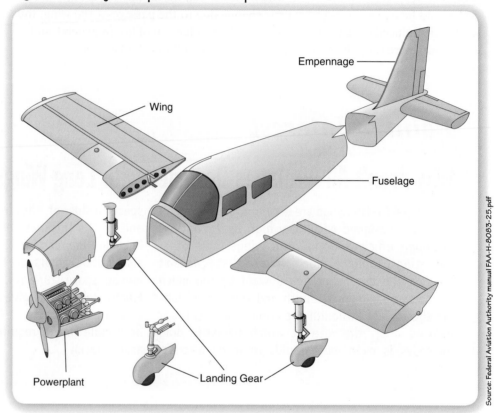

Source: Federal Aviation Authority manual FAA-H-8083-25.pdf

Most small training airplanes use an internal combustion, gasoline-fueled, four-cylinder, horizontally opposed engine, and a propeller to transform the chemical energy of the fuel into the mechanical energy of a rotating propeller. Larger aircraft frequently use jet engines that also burn a chemical fuel to create the mechanical energy that drives exhaust gases, a large fan, or a propeller to create thrust. More exotic powerplants include rockets, electrical motors, compressed gases, or twisted rubber bands (for very small, unpiloted aircraft).

Landing Gear

The largest forces experienced during a typical flight occur when the aircraft is brought back to Earth for landing. The landing gear's primary purpose is to absorb the large forces of landing without causing injury to people or damage to the aircraft and cargo that are on board. A secondary function is to reduce the frictional force with the ground to allow faster acceleration to takeoff speed and to reduce damage to the landing surface. A third function is to improve the ground handling of the aircraft during taxing. For some aircraft, the landing gear also serves as the means to attach to takeoff (catapult) and landing (wires) aids.

Fuselage

Flying isn't very useful unless you can take people, baggage, cargo, and equipment up with the aircraft. The fuselage provides the mounting and containing structure for all of these as well as for the other components of a typical aircraft. A few flying wings and other exotic designs integrate the fuselage's features directly into the wing itself.

Empennage

To improve the natural stability of an airplane and to reduce the size of the surfaces required to create sufficient forces for stabilization and maneuvering, most airplanes place the horizontal stabilizer, elevator, vertical stabilizer, and rudder behind the wing at the rear of the fuselage. This assembly of parts is collectively referred to as the empennage, and it creates the stabilization and control forces required to maneuver the aircraft.

Wings

The airplane's wings produce the aerodynamic force required to overcome weight and can contain control surfaces to maneuver the aircraft. Lift can be produced by other components of the airplane. Wings are typically mounted high on the fuselage for training and rough field airplanes and at mid- or low mounting points on the fuselage for more aerobatic aircraft. Mounting the horizontal stabilizer in front of the main wing is referred to as a canard design.

Wings are almost always attached to the fuselage of the aircraft at a slight angle to give the wing some AOA when the fuselage is horizontal and level in pitch. The amount of angle of incidence that should be built in is determined for cruising flight speeds and configuration, which represents the phase of flight in which an aircraft will spend most of its flight hours. This greatly enhances the comfort of crew and passengers in the aircraft. For aerobatic aircraft, there is seldom any angle of incidence because the aircraft spends a significant amount of its time in inverted flight.

WIND TUNNELS

A wind tunnel is a very useful device for testing the designs of an aerospace engineer. The use of a tunnel allows the direct measurement of forces as well as the visualization of airflow over surfaces. In addition, it frequently reveals unforeseen challenges to the vehicle performing as it was intended to perform.

Confirming Performance with Physical Testing

Wind tunnels are used to confirm the performance characteristics that were predicted during the design phase of a project. Wind tunnel models are created so that the full range of undesirable interactions between the complex flows around an aircraft can be uncovered and remedied prior to or during flight testing (see Figure 3-22). In some cases, wind tunnel testing is used to investigate aviation accidents and uncover root causes so that corrective measures can be developed to prevent further loss of life and property. To this end, wind tunnels come in all sizes and configurations to allow testing from subsonic to supersonic speeds.

Figure 3-22 Model of X-33 in the wind tunnel.

Courtesy of NASA Langley Research Center

A wind tunnel is made up of a few critical components. The working region of a wind tunnel is called the test section. This part of the tunnel contains all of the apparatus and instrumentation required to mount shapes in the tunnel and collect data on the shape's performance. The test section is frequently clear for smaller tunnels or has cameras mounted in it to allow for observation of flow visualization through surface tufts, smoke, or beads in the airflow. Modern tunnels collect numerical data in digital form so that it can be visualized and manipulated by computer.

To function properly, a wind tunnel requires a few additional components to ensure that air is brought into the test section with uniform velocity and very little turbulence. Wind tunnels also require a way to effectively get rid of, or recycle, the air leaving the test section so that there isn't a pressure buildup behind the shape being tested.

Your Turn

Wind tunnels come in all sizes and designs. Do a web search and find as many variations, sizes, and types of components as you can. Compare the list to the list of a classmate.

COMPUTATIONAL FLUID DYNAMICS (CFD)

The most high-tech aerospace engineering now makes use of a technique called **computational fluid dynamics (CFD)**. The CFD method uses powerful computers to carry out the billions of calculations required to simulate the full outcome of the equations governing fluid flow (see Figure 3-23). It wasn't until the 1960s that computers could produce three-dimensional estimates of performance based on much simplified equations. Throughout the 1970s and 1980s, computers became capable of carrying out the calculations using equations that more closely matched the actual characteristics of fluid flow, which allowed ever more accurate predictions of flight performance. However, designs continue to be prototyped, wind tunnel tested, and flight tested prior to manufacturing to ensure that the final product is safe and controllable.

Today, computers have become so powerful that aircraft can be designed, "built," and tested completely in the digital realm before the first physical prototype is ever constructed for flight testing. This method remains an expensive option for routine design tasks, however, it is more affordable than physical prototyping and testing for extremely high-performance and unique aircraft design challenges.

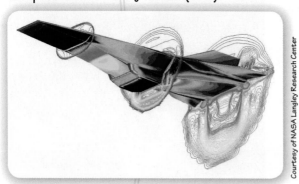

Figure 3-23 Hyper-X modeled using Computational Fluid Dynamics (CFD).

Courtesy of NASA Langley Research Center

Arrived at Destination

SUMMARY

■ Aircraft are flown by managing the magnitude and direction the four forces: lift, weight, thrust, and drag.

■ Lift is the result of deflecting air in a new direction and changing its pressure through an increase in its flow velocity.

■ Drag is the result of any two substances interacting. For an airplane, drag is induced by the production of lift and a result of parasitic frictional forces due to the aircraft's movement through the air.

■ Airfoils are specially shaped to produce lift without significant drag by guiding the airflow smoothly around the wing. Airfoils can be symmetrical for aircraft that spend a lot of time inverted but are more commonly asymmetrical with a more curved upper than lower surface.

■ Airplanes are constructed so that lift forces are generated and transferred to the fuselage, and so that crew, passengers, fuel, and baggage can be transported through all phases of flight.

■ Wind tunnels and computational fluid dynamics provide the means to test aircraft designs to confirm performance and discover undesirable traits prior to actual flight testing.

BRING IT HOME

1. Describe the changes that occur in the four forces during a typical flight.
2. Consider how the handles and grips on commercial products ensure that the forces applied to them lead to straight line or rotational motion.
3. Select a nonaviation sport or activity, and describe how drag reduction is accomplished to enhance performance.
4. Find a snapshot of an airplane on the Internet, and describe fully the visible and unseen components of the aircraft and their functions.

EXTRA MILE

1. Construct a balsa wood model such as those sold by the Guillow's company.
2. Create a box fan wind tunnel, and develop a way to measure both lift and drag relatively accurately.

CHAPTER 4
Flight Dynamics

GPS DELUXE

START LOCATION	DISTANCE	END LOCATION

Menu

Before You Begin

Think about these questions as you consider the concepts in this chapter:

1. How is stability different from maneuverability?

2. What is the center of gravity (c.g.) and how does it relate to flying an airplane?

3. How do pilots and engineers ensure that an aircraft is properly loaded for takeoff?

4. What cockpit controls are used to control the aircraft during flight?

5. How do the control surfaces on an airplane move to maneuver the airplane?

6. Can an airplane be designed to be inherently stable in flight?

7. What are the major phases of a typical flight?

Ask pilots about their first few hours of flight instruction, and you are likely to hear about the excitement of taking control of the airplane for the first time. After a few hours of instruction, pilots are confident in their ability to move controls in the cockpit, enabling the airplane to change course, climb or descend, or roll up on a wingtip to look down on their house. Almost all pilots also comment on their surprise at how easy it was to fly the airplane.

During any particular flight, an airplane will spend most of its time cruising in straight and level unaccelerated flight. During this phase of flight, the four forces of lift, weight, thrust, and drag are in equilibrium with each other so that the aircraft moves with a constant velocity. For our new student pilot, a well-designed training aircraft is a very stable airplane that will, on its own, tend to return to unaccelerated flight. To be controllable, the trainer is designed to create small imbalances between lift and weight, and between thrust and drag that cause the aircraft to maneuver through the air. Thus, the aircraft has to be both stable and maneuverable; these two opposing goals for the design are essential if the aircraft is going to be both safe and useful.

This chapter focuses on how the aircraft and pilot work together to change the four forces of flight to accomplish the goal of stability and maneuverability in flight. By the end of the chapter, you will understand the mechanisms on the airplane, the actions of the pilot, and the resulting changes in the four forces that make accomplishing the goals of stable and maneuverable flight possible.

Courtesy of NASA

NASA's F-15 Advanced Control Technology for Integrated Vehicles (ACTIVE) aircraft achieved its first supersonic yaw vectoring flight at Dryden Flight Research Center, Edwards, California, on April 24, 1996.

FORCE AND MOMENT BALANCE

The simplest of all forces to understand is weight. We experience it every day, from walking to class to playing sports. Add more objects to a backpack, and it not only becomes heavier but you also have to lean farther forward to maintain your balance. This experiential knowledge of gravity, and its consequence of weight, applies to airplanes as well.

Weight and Balance

The complicating factor for an airplane is that it is flying with no visible means of support. Disconnected from the earth, we have to wonder where the airplane can balance. The answer is found in the concept of center of gravity (c.g.). Pick up a child's teddy bear by the ear and the body swings due to gravity to place itself below the point of support. Grab it again by an arm and it swings around again to a stable position. The entire bear is acting as if its mass was concentrated at one location, which is always pulled below its support. This location is the center of gravity of the object (recall this concept from Chapter 3). In the same way that all of the lift being produced at millions of locations on the wing can be summarized by one lift vector, so too the c.g. represents the place at which we can act as if all of the airplanes weight is being exerted.

Your Turn

Find a number of differently shaped objects in your house, and try the experiment described for the teddy bear. Can you find the location of the c.g.? What happens if you move your fingers to the c.g. to hold the object?

The c.g. for an airplane can change any time weight is added or removed from the aircraft. This is important to remember when baggage, passengers, and fuel are routinely added and removed from the airplane or are shifted to a new location in the aircraft.

Understanding c.g. allows us to consider lift and weight as two vectors. If the two are equal in strength and are applied at the same location, the aircraft will be unaccelerated in vertical flight and will continue what it is doing regardless of whether that is a climb, descent, or cruising flight.

Initial Loading

Move the c.g. just forward of the lift's center of pressure and two forces will still be of equal magnitude, but the aircraft will not stay in its current orientation. Gravity attempts to pull the c.g. under the lift vector by pitching the nose downward. Put the c.g. behind the lift vector, and the aircraft pitches upward. To compensate for this effect, the pilot uses movable control surfaces to create aerodynamic forces to balance out the torque and hold the nose at the necessary pitch.

When the airplane was designed, the engineers had in mind the ability to carry a specific amount of weight at known locations in the airplane, for example, fuel in fuel tanks and baggage in the cargo hold. Because the amount of aerodynamic force a control can create is limited by the lift it can produce (area, coefficient of lift [C_L], velocity, air density), we must keep the c.g. within a fairly narrow area in the airplane for it to remain controllable. To ensure that the c.g. is within the allowable limits, pilots use mathematical equations, charts, or computer software to calculate the location of the c.g. before every flight.

Weight and Balance Calculations

Ensuring that an airplane is ready for flight involves confirming three essentials prior to takeoff. The weight of the aircraft must be sufficiently less than the maximum the aircraft is capable of lifting to provide sufficient net force to maintain a rate of climb in line with the aircraft's mission; the torque created by the objects loaded on the aircraft must be small enough that it can be balanced by normal control surface movements; and the load must be secure so that no object can shift far enough to cause the aircraft to become uncontrollable.

The strength of the torque created by each object on the airplane is referred to as its **moment**. Calculating the moment is simple multiplication of the object's weight times its distance away from some location used as the reference datum. Many manufacturers use the tip of the airplane's nose as the location for the reference datum. The following examples illustrate how weight and balance work.

Example 1

We already know the seesaw balances, so let's use the situation to learn some math (see Figure 4-1). If we put the reference datum at the pivot point for the seesaw, then it is symmetrical, and we can act as if the device itself doesn't exist.

Figure 4-1 A seesaw can show how weight and balance work.

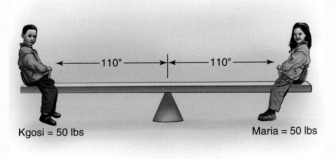

Kgosi = 50 lbs   Maria = 50 lbs

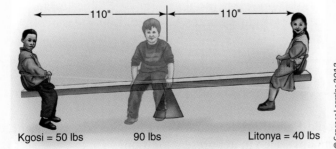

Kgosi = 50 lbs   90 lbs   Litonya = 40 lbs

© Cengage Learning 2012

Equation 1:   *Torque (or Moment) = Weight × Distance*

Moment due to Maria: 50 *lbs* × 110 *inches* = 5500 *lb · inches* Clockwise

Moment due to Kgosi: 50 *lbs* × 110 *inches* = 5500 *lb · inches*
Counterclockwise

The seesaw is balanced.

Example 2

Maria trades places with Litonya who only weighs 40 lbs. We know from experience that Kgosi's weight will lift Litonya into the air, so this is an excellent opportunity to show that the math confirms our experience.

Moment due to Litonya: 40 *lbs* × 110 *inches* = 4400 *lb · inches* Clockwise

Moment due to Kgosi: 50 *lbs* × 110 *inches* × 5500 *lb · inches*
Counterclockwise

The seesaw is not balanced. To find the amount of unbalanced moment that causes the seesaw to maneuver, we need to sum the two moments. This also means that we need to declare one of the two directions to be negative. In this example, counterclockwise is declared the negative direction and the individual moments are added together to find the total moment:

$$4400 \text{ } lb \cdot inches + (-5500 \text{ } lb \cdot inches) = -1100 \text{ } lb \cdot inches$$

Because the result is negative, the seesaw rotates counterclockwise, exactly matching with our experience. To make this a more complete example, we will find the new location of the c.g. for Kgosi and Litonya by taking the total moment (notice that the math for finding the total moment was shown as adding a negative number) and dividing this by the total weight of the system. From equation 1, and a little algebra, this will solve for the equivalent arm, or distance to the c.g., for the system.

$$Distance \text{ } to \text{ } CG = \frac{Total \text{ } Moment}{Total \text{ } Weight} = \frac{-1100 \text{ } lb \cdot inches}{(50 \text{ } lbs + 40 \text{ } lbs)}$$

$$= \frac{-1100 \text{ } lb \cdot inches}{90 \text{ } lbs} = -12.2 \text{ } inches$$

Notice that the c.g. shifted to the left. The two people act exactly like one 90 lb person sitting 12.2 inches to the left of center on the seesaw.

Challenge Question

Where would Kgosi have to sit for the seesaw to become balanced again if Litonya remains seated in her present location?

Example 3

Consider an airplane capable of creating 2,000 lbs of lift and that remains controllable if the c.g. is located between 45 and 49 inches behind the reference datum at the tip of the nose spinner (see Figure 4-2).

Figure 4-2 **Typical weights for a small aircraft with pilot and instructor on board.**

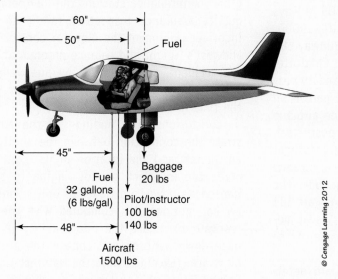

Consider the moment created by a few objects in our example airplane:

▶ The airplane itself weighs 1,500 lbs, and its c.g. is located 48 inches from the nose of the airplane.

▶ The aircraft is being flown by 17-year-old student pilot Qiaohui who weighs 100 lbs and is seated next to her flight instructor who weighs 140 lbs. Both are seated 50 inches from the airplanes nose.

▶ The aircraft is loaded with 32 gallons of fuel (6 lbs per gallon) in tanks centered 45 inches from the reference datum.

▶ The pilot's bag containing books, charts, flashlights, and other gear weighs 20 lbs and is located in baggage area one located 60 inches from the reference datum.

Is this aircraft safely loaded for flight?

To answer this question, we need to solve for the total weight, total moment, and location of the c.g. as described.

$$Total\ weight = Aircraft + Passengers + Fuel + Baggage$$

$$Total\ weight = 1500\ lbs + (100\ lbs + 140\ lbs) + \left(32\ gallons \times 6\frac{lbs}{gallon}\right)$$

$$+\ 20\ lbs = 1952\ lbs$$

Moment due to aircraft: $1500\ lbs \times 48\ inches = 72000\ lb \cdot inches$

Moment due to passengers: $(100 + 140)lbs \times 50\ inches = 12000\ lb \cdot inches$

Case Study ⟫⟩→

AIR MIDWEST ACCIDENT CLAIMS LIVES

National Transportation Safety Board Executive Summary

On January 8, 2003, about 08:47:28 eastern standard time, Air Midwest (doing business as US Airways Express) flight 5481, a Raytheon (Beechcraft) 1900D, N233YV, crashed shortly after takeoff from runway 18R at Charlotte-Douglas International Airport, Charlotte, North Carolina. The 2 flight crew members and 19 passengers aboard the airplane were killed, 1 person on the ground received minor injuries, and the airplane was destroyed by impact forces and a post crash fire. Flight 5481 was a regularly scheduled passenger flight to Greenville-Spartanburg International Airport, Greer, South Carolina, and was operating under the provisions of 14 *Code of Federal Regulations* Part 121 on an instrument flight rules flight plan. Visual meteorological conditions prevailed at the time of the accident.

The National Transportation Safety Board determined that the probable cause of this accident was the airplane's loss of pitch control during takeoff. The loss of pitch control resulted from the incorrect rigging of the elevator control system and the airplane's aft c.g., which was substantially aft of the certified c.g. limit. Contributing to the cause of the accident were Air Midwest's and the Federal Aviation Administration's (FAA) lack of oversight of the work being performed at the Huntington, West Virginia, maintenance station, the Raytheon Aerospace quality assurance inspector's failure to detect the incorrect rigging of the elevator control system, Air Midwest's weight and balance program at the time of the accident, and the FAA's average weight assumptions in its weight and balance program guidance at the time of the accident.

Conclusions: The findings of the NTSB demonstrate the connection between the ability to control an aircraft and the location of its c.g.. In this aircraft, the incorrect control rigging to the elevator limited its downward travel to only about half of its normal motion. This combined with the c.g.'s shift rearward to reduce the pilots ability to maintain a nose down attitude and control the aircraft's angle of attack. (Excepted from the National Transportation Safety Board, NTSC Publication Number AAR-04/01 www.ntsb.gov/publictn/2004/AAR0401.htm)

Many Federal Aviation Administration's regulations are the result of analyzing accidents of this type and identifying ways to prevent the accident from occurring again.

Moment due to fuel:

$$\left(32 \text{ gallons} \times 6\frac{lbs}{gallon}\right) \times 45 \text{ inches}$$
$$= 8640 \text{ } lb \cdot inches$$

Moment due to baggage: $20 \text{ } lbs \times 60 \text{ } inches = 1200 \text{ } lb \cdot inches$

Because all the moments are behind the reference datum, we leave them all positive when we find the total moment.

$$Total \text{ } Moment = (72000 + 12000 + 8640 + 1200)lb \cdot inches$$
$$= 93840 \text{ } lb \cdot inches$$

The c.g. will be located a distance from the reference datum equal to the total moment divided by the total weight of the system.

$$Distance \text{ } to \text{ } the \text{ } CG = \frac{93840 \text{ } lb \cdot inches}{1952 \text{ } lbs} = 48.1 \text{ inches}$$

This airplane is safe for flight. You are cleared for takeoff!

Normal Flight

During a typical flight in a small airplane, the only change in weight and balance is the consumption of fuel during the flight. For this reason, fuel tanks are almost always located very close to the center of lift so that the location of the c.g. doesn't shift very much as fuel is burned. In a large commercial jet, there can be many fuel tanks located throughout the airframe. If the fuel tanks are not located close to the center of lift, then fuel management is very important to make certain that the aircraft's c.g. does not shift too far during flight.

Aircraft can also routinely experience a change in weight and balance as passengers move around the cabin, baggage is moved, ordinance is dropped, or jumpers leave the aircraft. Anything that changes the airplane's weight or the balance of forces will affect its flight performance and stability in flight.

Weight Shift Vehicles

Since the dawn of aviation, aerial vehicles have been designed to shift the c.g. to control the roll and pitch of the aircraft. Today, there is a resurgence of weight-shift aviation as many ultra-light airplanes and powered parachutes are designed to use this simple and effective means of controlling a lightweight aircraft.

AERODYNAMIC CONTROL SURFACES

The most common way to maneuver an aircraft is by changing the magnitude or direction of the lift produced by a surface on the aircraft. The resulting change in force causes the aircraft to change its orientation in space and, as a result, its motion through the air.

Axes of Rotation and Motion

The three primary axes through an aircraft define the orientation of the aircraft and its motion through space. All three lines pass through the location of the aircraft's c.g. so that forces applied anywhere other than directly through the c.g. will cause the airplane to rotate around one or more of the axes of rotation. The three axes are called the longitudinal axis (in the direction of flight from nose to tail), lateral axis (perpendicular to the direction of flight from wingtip to wingtip), and vertical axis (perpendicular to the other two axes) (see Figure 4-3). Rotation around each axis results in changes in roll, pitch, and yaw, respectively.

Figure 4-3 Axis of rotation and direction of motion.

Source: http://www.faa.gov.library/manuals/aviation/pilot_handbook/media/PHAK%20-%20Chapter%2004.pdf

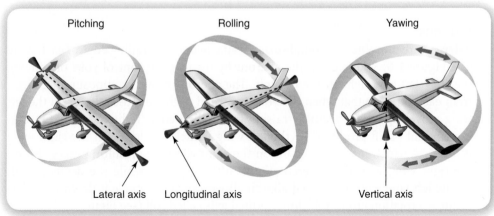

Control Surfaces

Control surfaces are movable parts of an airplane designed to allow the pilot to control lift to maintain or change the motion of the airplane in pitch, roll, and yaw. Aircraft typically are designed to use an elevator, ailerons, a rudder, flaps, and trim tabs.

> **Control surface:**
>
> movable parts of an airplane designed to allow the pilot to control lift to maintain or change the motion of the airplane in pitch, roll, and yaw.

Elevator

The **elevator** is a movable horizontal surface attached to the trailing edge of the horizontal stabilizer on most aircraft (see Figure 4-4). During routine cruising flight, the elevator is designed so that it is aligned with the horizontal stabilizer to minimize drag. From the cockpit, the pilot can push the yoke or control stick forward to cause the elevator to move downward. This movement effectively moves the stabilizer's trailing edge downward, increasing its angle of attack (AOA), which increases the lift it produces. The result is rotation around the lateral axis called pitch (see Figure 4-5).

Movement of the elevator downward attempts to pitch the nose of the airplane downward from the pilot's perspective. Pulling the yoke or control stick backward accomplishes the opposite result and pitches the nose upward as seen by the pilot. In aerobatic flight, the direction of these relative motions is important because a pilot would have to push the stick forward in inverted flight to pitch the nose higher above the horizon.

Figure 4-4 *Location of the elevator.*

© Image by X-Plane

Figure 4-5 *Movement of the elevator, shown with elevator centered, deflected downward, and deflected upward.*

© Image by X-Plane

Ailerons

Ailerons are movable horizontal surfaces attached to the trailing edge of each wing (see Figure 4-6). Pilots move the ailerons by turning the control yoke or moving the stick left and right in the cockpit. The ailerons move in opposite directions from each other so that when the yoke is turned to the left, the aileron on the left wing moves upward, effectively raising the trailing edge, reducing the AOA and the amount of lift produced by the left wing while the aileron on the right wing is lowered, which increases the lift on the right wing. The resulting excess lift on the right wing with the lowered aileron lifts it upward while the decreased lift on the left wing with the raised aileron allows gravity to move it downward. The result is a rotation around the longitudinal axis called roll (see Figure 4-7).

Figure 4-6 *Location of the ailerons.*

Figure 4-7 *Movement of the ailerons, shown with the control yoke centered, deflected counter clockwise and deflected clockwise.*

Ailerons can run full span from the root to the tip of each wing but more typically are found on the outer half of each wing's span. Remember that moments are the product of a force and a distance from the center of rotation. When ailerons deflect, they produce a change in lift. If the same lift force is created far away from the airplane's c.g., then the resulting moment is very large, whereas a lift force close to the c.g. creates a very small moment. Because creating lift also creates drag, it is much more effective to put the ailerons as far out on the wings as possible to minimize the deflection required to maneuver the airplane (see Figure 4-8). If an airplane needs more control in roll, or to roll more quickly, the ailerons can be made larger both in chord and in span. In the extreme case of many modern fighter jets, the combined horizontal stabilizer and elevator (stabilator) move in opposite directions on each side of the aircraft to improve the roll rate of the aircraft.

Creating aerodynamic lift always creates drag as well. When the ailerons are deflected, we are increasing the lift, and therefore the drag, on only one wing panel. On the other side of the aircraft, the other wing is experiencing decreased lift and drag. The result is that the wing that is moving upward is also moving less quickly forward when compared to the wing that is descending. The difference in drag and its resulting motion is referred to as adverse yaw. Countering adverse yaw is an important part of aircraft design.

Figure 4-8 *Cessna 172SP in a left turn.*

Rudder

The **rudder** is a movable surface attached to the trailing edge of the vertical stabilizer. From the cockpit, the pilot presses on pedals on the floor to move the rudder. During

normal flight, the rudder is aligned with the stabilizer to minimize drag. Stepping on the left rudder pedal moves the rudder to the left. The result is the creation of an AOA that creates a lift force to the right and a motion of the nose of the airplane to the left. The rudder controls yaw, which refers to the forces that cause rotation around the vertical axis (see Figure 4-9). Coordinating the amount of rudder displacement with aileron deflection is an important part of countering adverse yaw (see Figure 4-10).

Figure 4-10 Location of the rudder pedals in the cockpit.
© Image by X-Plane

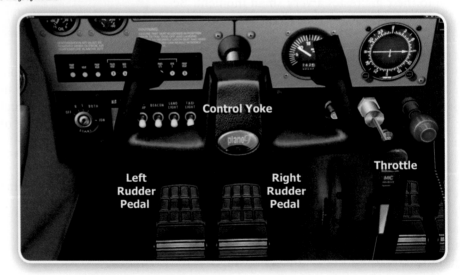

Trim Tabs

Frequently during flight, the pilot is required to maintain pressure on the yoke or rudder pedals to hold the control surfaces in the desired position. This can be very tiring during even short flights. On most airplanes, small movable surfaces on the control surfaces called **trim tabs** (see Figure 4-11) are used to create a lift force that holds the control surface in position. So if we need help holding the elevator in a downward deflection (to help hold the nose down in pitch), we can move the elevator trim tab upward. Trim tabs may be controllable from the cockpit, or they may be simple sheets of metal that are bent into position while on the ground.

> **Trim tabs:**
> small movable control surfaces attached to a larger control surface such as the rudder, elevator, or ailerons so that it can create an aerodynamic force to hold the larger control surface in a given position without any control pressure in the cockpit.

Figure 4-11 *Components of the empennage of a typical airplane shown on a P-51 Mustang.*

Vertical Stabilizer

Rudder Trim Tab

414320

L2

Horizontal Stabilizer

Elevator

Elevator Trim Tab

Rudder

Photo: © Ben Senson

Physics Connection

Movements, Simple Machines, Levers

Recall that the balance point on an aircraft is the c.g.. When a force acts on an object in the same way the control surfaces act on an aircraft, it might help to visualize the interaction as second-class levers and torques. Remember torque (τ) is defined as a force applied some distance from a fulcrum: $\tau = F \times d$. How does the location of a control surface affect its performance? If you could place ailerons farther out on the wing, could they be smaller?

What about elevators and rudders? According to our formula, smaller forces farther away from the fulcrum can produce the same amount of torque. If we cannot have a large distance between the fulcrum and the force, then we must compensate by using larger control surfaces. If you understand this concept and the purpose of ailerons, you should never get them confused with flaps again.

Flaps

Flaps are movable surfaces mounted to the trailing edge of each wing (see Figure 4-12). The flaps move together in the downward direction. As the flaps are deflected, the trailing edge is lowered, and the AOA is increased so that the wings will create more lift and more drag. When deflected only a little, there will be a large gain in lift with little gain in drag. As the flaps are increasingly deflected, there is less increase in lift while the drag increases dramatically. Many aircraft use some flap deflection during takeoff and full flap deflection during landing.

The advantage of using flaps to change the AOA of the wing extends beyond creating more lift. Dual benefits exist because the airplane can maintain a more level attitude in pitch, which greatly improves the pilot's view of the runway, and

Figure 4-12 Location of the flaps shown on Cessna 172SP in flight with flaps deployed.

Image by X-Plane

Figure 4-13 Flaps are movable control surfaces that increase lift and drag by deflecting downward and/or extending into the airstream at the wing's trailing edge.

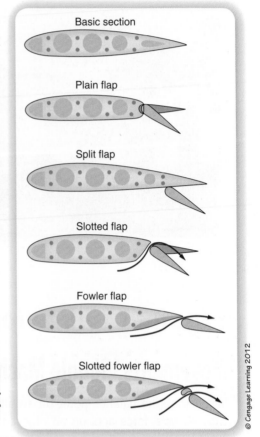

Basic section

Plain flap

Split flap

Slotted flap

Fowler flap

Slotted fowler flap

© Cengage Learning 2012

the airplane can approach at a slower speed while descending at the same rate during the approach for landing.

Because both flaps move downward together, there is no net moment created along the longitudinal axis, and the aircraft will not experience roll due to use of the flaps.

Flaps systems can be very simple or complex and can be combined with full span ailerons in flaperons to provide both functions with one control surface (see Figure 4-13).

Other Control Surfaces (LE Slats, Spoilers, Wing Warping)

Alternative lift control devices include leading edge slats (LE slats), spoilers, as well as the Wright brothers' wing warping technique. **Leading edge slats** are devices that can be extended forward of the leading edge to guide air's passage over the wing. This allows the wing to be flown at higher AOAs as well as increasing the wing's planform area. **Spoilers** are movable surfaces that can be extended upward into the airflow as it passes over the airfoil. They create significant turbulence, which spoils the lift effects of the wing on that side while increasing drag. Thus, when a spoiler is deployed on only one wing, it can act similar to ailerons to induce roll. **Wing warping** twists the structure of the entire wing to create a different AOA on each wing. Modern techniques for flexible wings are under development for very small aircraft.

TURNING FLIGHT

A typical turn is initiated with ailerons (see Figure 4-14). As the yoke is turned in the direction of the intended turn, the outside wing rolls upward and moves backward relative to the fuselage due to adverse yaw. The pilot applies pressure to the inside rudder pedal to induce a lift force at the tail that induces a yaw that keeps the nose aligned with the current direction of flight.

Figure 4-14 Lift vector for an airplane in straight and level, unaccelerated flight.

Figure 4-15 Lift vector for an airplane rolled into a left turn. Notice that the vertical component of this lift vector would be reduced in magnitude and the airplane would descend.

As the aircraft rolls, the lift vector produced by the wings is tilted in the direction of the intended turn. If the lift vector is broken into vertical and horizontal components, the horizontal component increases in value, pulls the aircraft to the side, and causes the aircraft to turn (see Figure 4-15). The vertical component of lift decreases as the aircraft rolls, which causes the aircraft to lose altitude unless back pressure is applied to the yoke to deflect the elevator upward. The elevator deflection results in a pitch up of the nose and an increase in total lift as the wings' AOA increases. The greater lift allows the aircraft to maintain altitude and execute the turn simultaneously.

In an extreme bank attitude and the resulting tight turn, the outer wing moves much faster through the air than the inner wing. The resulting increase in lift on the outer wing panel can cause a continued roll into the turn called over-banking (see Figure 4-16). Pilots routinely make minor adjustments to aileron, rudder, and elevator position to initiate and roll out of turns.

Figure 4-16 Lift vector for an airplane rolled into a left turn. Back pressure on the stick or yoke induces a larger angle of attack, and more total lift is produced increasing the vertical lift back to its original value. The aircraft holds altitude while the horizontal component of lift pulls the aircraft through the turn.

STABILITY IN FLIGHT

Stability is the natural tendency of an object to maintain and return to its original orientation. For an airplane, the desired attitude is straight and level flight. To be easy and safe to operate, an airplane needs to be stable in roll, pitch, and yaw. Most recreational aircraft are designed to be inherently stable. Most aerobatic or fighter aircraft are inherently unstable. Why would engineers want an aircraft that is unstable?

Stability:
the natural tendency of an object to maintain and return to its original orientation.

Roll

Stability in roll is referred to as lateral stability and is influenced by four major design characteristics of the airplane. Aerospace engineers have a lot of control over the lateral stability of an airplane design as they determine the location of objects in the aircraft, the dihedral and sweepback of the wings, and the area and location of the aircraft's vertical surfaces.

Weight distribution is primarily affected by the location of fuel tanks and a pilot's ability to monitor fuel levels to maintain an equal amount of fuel on each side of the longitudinal axis. Many aviation accidents are the result of improperly monitoring and managing fuel onboard the aircraft.

Dihedral is the upward angle at which a wing is mounted to the airplane fuselage when compared to the horizontal (see Figure 4-17a). Dihedral is typically very small for high wing aircraft but can be very significant for low wing aircraft or nonexistent for aerobatic aircraft. Dihedral improves lateral stability during roll maneuvers by increasing the AOA of the descending wing and decreasing the AOA of the rising wing. This creates a difference in the lift produced by each wing that tends to return the aircraft to level flight (see Figure 4-17b).

Sweepback is the angle at which the wing's leading edge is mounted to the fuselage compared to the lateral axis (see Figure 4-18). When an airplane rolls, the aircraft starts to descend and move sideways. This sideslip makes the descending wing more perpendicular to the relative wind and increases its lift compared to that of the ascending wing. This induces the airplane's return to level flight.

Pitch

Pitch stability depends heavily on weight and balance, as previously discussed, and on the angle at which the horizontal stabilizer is mounted on the fuselage. The horizontal stabilizer on most aircraft is mounted with a slightly negative AOA. If the nose drops, the airplane increases its velocity, which increases the downward aerodynamic force at the tail, which pushes the nose upward to its original attitude. Likewise, if the nose is pitched up, the airplane slows, the downward force at the tail decreases, and the nose drops back to its original orientation.

Yaw

Stability around the vertical axis is referred to as directional stability. Directional stability is strongly affected by the keel effect, wing sweep, and drag-inducing devices on the wings. The keel effect is the result of having more surface area exposed behind than in front of the c.g. when the aircraft is not aligned with the direction of flight. This creates a larger torque pushing toward alignment with the flight direction than that pushing away from alignment.

Figure 4-17a Dihedral is the result of mounting the wings with the wingtips elevated above the wing root.

© Cengage Learning 2012

Figure 4-17b *Effect of dihedral during level flight and when the aircraft is in a rolled attitude.*

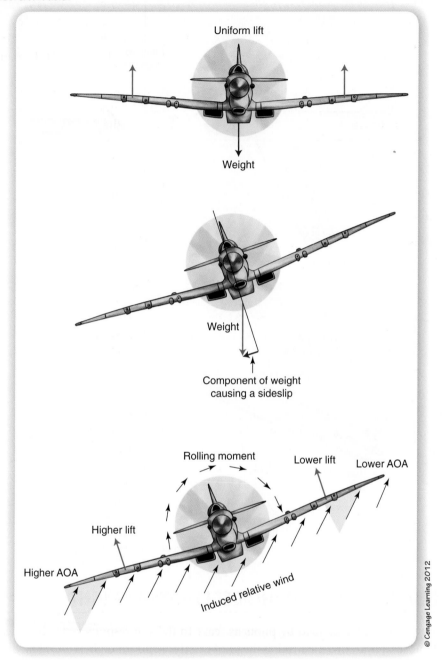

Wing sweep impacts directional stability through the creation of lift and induced drag. For the same reason that wing sweep improves roll stability, the wing that is more perpendicular to the direction of flight will have more drag exerted on it. This in turn causes the nose to yaw back into alignment with the direction of flight.

Directional stability is a real challenge for aircraft that lack a traditional fuselage and empennage. In particular, flying wing designs often use wing sweep in combination with computer-controlled, drag-inducing control surfaces near the wingtips to control yaw.

Figure 4-18 Wing sweep and roll stability.

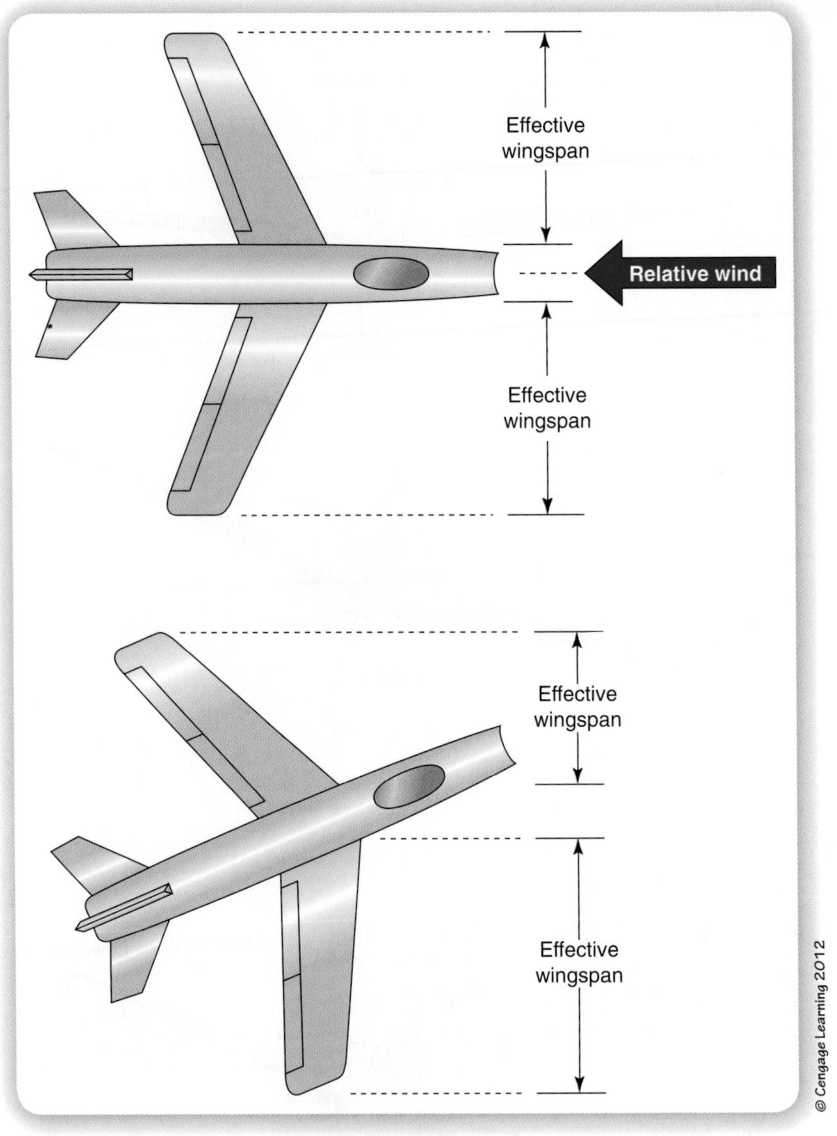

© Cengage Learning 2012

EQUILIBRIUM AND UNACCELERATED FLIGHT

Aircraft that are described by pilots as "easy to fly" are responsive to their needs. For most aircraft, this means that the pilot can fly a smooth flight without having to make constant, significant changes to control positions or maintain a force on any controls. A properly designed and trimmed aircraft satisfies these demands.

Phases of Flight

All aircraft have to execute the major phases of flight to accomplish meaningful tasks. Airplanes need to taxi, take off, climb, cruise, descend, and land safely. During each of these phases of flight, the four forces of lift, weight, thrust, and drag are adjusted to produce the desired results (see Figure 4-19).

Taxi and Ground Maneuvers

To taxi an aircraft, we want to produce either no lift or negative lift to maintain friction between the aircraft's tires and the airport surface. In addition, when the

Figure 4-19 *Phases of flight and safety.*

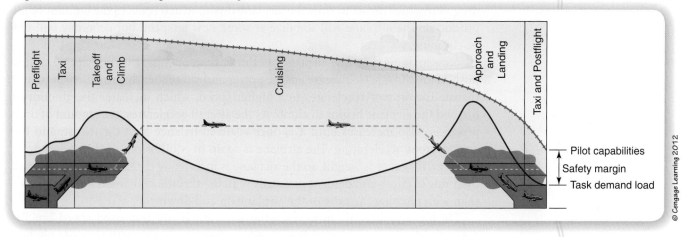

wind gusts, we want to make certain the aircraft isn't lifted off the ground and flipped. This means that during taxi operations, pilots hold the yoke forward and into the wind to move the elevator downward and raise the aileron on the side the wind is gusting from.

Takeoff

After we reach the runway, it is time for takeoff. During slow-airspeed, high-power situations, an airplane will have a strong left-turning tendency due to three factors. From the pilot's perspective, the engine spins the propeller in a clockwise direction. This creates a counter-torque that attempts to roll the aircraft to the left. In addition, after the aircraft pitches up to increase its AOA for climbing, the descending propeller blade will have a higher AOA than the ascending blade due to the relative wind and pitch angle. This means that the propeller produces more "lift" or thrust on the right side of its travel than on the left side. This "P-factor" tends to yaw the nose to the left. Finally, the spinning of the propeller causes the air to flow around, as well as past, the fuselage in a spiraling slipstream. This slipstream tends to strike the left side of the vertical stabilizer, which also induces a yaw to the left. During basic flight instruction, student pilots frequently hear the instructor's call to hold more right rudder to hold the runway center line. Larger aircraft allow the pilot to change the rudder trim setting so that the rudder holds itself to the right with aerodynamic forces rather than pilot-generated forces.

Liftoff

The moment of liftoff is the beginning of a climb that requires a gain in potential gravitational energy as we lift the entire aircraft farther and farther from Earth's surface. However, most climbs are executed while maintaining a constant speed, which means that the classic tradeoff of kinetic and potential energy that is emphasized in basic physics courses isn't the source for the energy to climb for most aircraft. Instead, an airplane climbs on excess thrust. The following example helps make this easier to understand.

Climbing

Imagine an airplane in straight and level, unaccelerated flight. Thrust exactly equals drag, so the aircraft does not speed up or slow down. Lift balances weight, so the aircraft is neither climbing nor descending. When we pull back on the yoke to lift the nose, the immediate change in AOA produces excess lift, and the aircraft begins to climb. However, producing more lift also produces more drag. The increased drag overcomes thrust to slow the airplane down, which reduces the amount of lift produced. The airplane will stabilize at some new airspeed but generally will not be climbing.

Now imagine the same airplane but rather than pitching the nose up, we simply push in the throttle to increase engine power and subsequently propeller speed and thrust. The aircraft accelerates to a higher speed, which increases the amount of lift, and the airplane begins to climb. As the aircraft accelerates, the amount of drag produced also increases until it is once again in balance with the thrust, and the aircraft stops accelerating. The airplane is again in equilibrium in thrust and drag; however, lift exceeds weight, so the airplane is in a steady climb. Real pilot's control their rate of climb and descent primarily with the throttle and power settings rather than with pitching the nose of the airplane up and down.

During a real climb, however, what we want is not increased speed but increased altitude. To achieve this goal, pilots trim the aircraft with the nose pitched up. Remember that unlike our first example, we are also increasing the throttle setting to overcome the increase in drag. By changing pitch and power together, we can transition into and maintain a constant speed climb. During takeoff, pilots also lower the flaps just a little. This increases the wings' effective AOA without pitching the nose up or inducing much drag while greatly improving the pilot's view over the nose of the aircraft.

It's important to note that gliders, which don't have an engine, can only achieve very temporary increases in altitude by exchanging airspeed for altitude and slowing down. Barring the existence of an outside source of energy, every time a glider attempts to climb, it achieves a lower and lower maximum altitude due to the loss of energy to drag. So how do gliders stay in the air so long if they are constantly descending?

Point of Interest

How Gliders Fly

Gliders tap into external energy sources such as a tow plane, a truck to pull them to the mountain top (or your legs!), stretched bungee cords or winches on the ground, and rising air currents due to the warm air rising in a thermal or orographic lifting as air flows over rising ground. Once in flight, gliders descend through the air, but if you are skilled, you may be able to climb by flying into air that is rising as a mass faster than your rate of descent within the air mass.

Cruising Flight

Cruising flight is the straight and level, unaccelerated flight intended to get you from point A to B as efficiently as possible. Most airplanes are designed for maximum efficiency during this phase of flight. To achieve cruising flight, a pilot needs to adjust the throttle to the engine's most efficient power setting and then trim the elevator to hold a pitch attitude that produces no change in altitude.

The faster an aircraft flies, the more lift is produced by velocity and the less AOA that is required to maintain altitude. Induced drag is significantly reduced at higher speeds at the same time that parasitic drag increases. The balance point at which the least amount of total drag is produced is the **maximum lift to drag ratio (L/D$_{Max}$)**. Flying at this airspeed maximizes the range of the aircraft and is important during engine out or low fuel emergencies. In the real world, we frequently want to travel faster than the L/D$_{Max}$ speed. With rising fuel prices, expect to see air travel take slightly longer to get to your destination as the amount of excess fuel is reduced that is burned to fly faster than this speed (see Figure 4-20).

Descending Flight

Descending flight is identical to climbing flight except the lift force is adjusted to be less than the airplane's weight so that gravity wins and the aircraft descends at constant speed. A critical moment in any descent is the moment when we arrive at Earth's surface and need to transition for landing. Landing the airplane is the most dangerous phase of flight. The reality is that pilots spend the entire flight making certain that the airplane doesn't collide with any other object, and then during the approach and landing, they intentionally set up an impact with the ground.

Configuring the airplane for landing generally requires slowing down the airplane. The slower the airplane is flying, the more time the pilot has to make corrections to the approach to landing and the smaller the forces are on the landing gear and flaps as they are extended from the aircraft. Remember that to slow down and maintain altitude, or slowly descend, we need to increase the pitch up attitude of the airplane. This is very unsafe because it lifts the nose of the airplane into the sightline of the pilot at the critical time the pilot needs to see the runway or approach lighting.

Dropping the flaps significantly during a landing allows the pilot to dramatically increase the wing's effective AOA while keeping the nose in a lower pitch attitude (see Figure 4-21). Flaps also solve the problem presented by the exchange of potential gravitational and kinetic energy. During a descent, the aircraft wants to speed up. Extended flaps help to reduce the airplane's speed by inducing a very large amount of drag when they are fully extended. For both of these reasons, pilots use nearly fully extended flaps during the landing phase of flight.

Landing

The transition to landing starts with flight into ground effect. Remember that both takeoffs and landings are affected by ground effect, which significantly reduces drag within a wing span distance of the ground. During the approach, the airplane is configured and maneuvered to have the minimum airspeed possible as the aircraft crosses the runway threshold and enters ground effect. At the last moment, the throttle is reduced as the nose is pitched up to produce the flare for landing that reduces the descent rate to zero as the wheels make contact with the surface. Many aircraft use brakes to reduce speed until it is safe to taxi the aircraft. Larger aircraft use thrust reversing where the powerplant's force is sent forward to slow the aircraft quickly. Some unusual conditions during landing may require the use of drag chutes to slow the aircraft's approach and rolling distance after takeoff. This is useful for very high-speed, low-drag aircraft with limited braking such as the Space Shuttle.

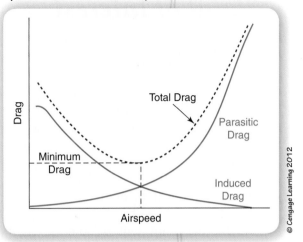

Figure 4-20 *Sources of drag for a typical wing operating at various airspeeds.*

© Cengage Learning 2012

Figure 4-21 *Approach lighting to Chicago O'Hare runway 32L.*

© Cengage Learning 2012

Point of Interest

The Carrier Landing

The most exotic landing is that of a Navy aircraft carrier landing, or "trap" (see Figure 4-22). Because the carrier's deck is so short, the only way to stop a landing aircraft on the deck is to have it catch, or trap, a wire that exerts a strong, calibrated, backward pull on the airplane to bring it to rest. The strength of this force is adjusted to match the mass and velocity of each aircraft as it lands. To improve the odds of catching the wire, the carrier steams directly into the wind, which reduces the airplane's relative approach speed, and the aircraft's descent rate is not reduced to zero so that the airplane "gently" crashes into the deck. Navy planes are built tough!

The problem is what happens when the airplane misses the wire and becomes a "bolter." In addition to the steps already described, to regain flight speed before running out of deck, the aircraft has to be at full throttle at the moment of touchdown. Add to this the rolling motion of the carrier's deck and the chance of inclement weather, a damaged aircraft, or low fuel, and you can appreciate that being a carrier-rated pilot is one of the most demanding jobs in aviation.

Figure 4-22 Flyover and landing of an FA-18 on a carrier deck.

Courtesy of United States Navy

SUMMARY

- Weight creates a moment due to its location in the aircraft.

- An improperly loaded aircraft can be uncontrollable from the cockpit.

- Weight and balance is the method used by pilots to confirm that their aircraft is properly loaded prior to takeoff. Their primary concerns are that the aircraft doesn't exceed its maximum allowable weight, that the c.g. is within allowable limits, and that the load is secured so that it cannot shift during the flight.

- Control surfaces create aerodynamic forces that cause rotation around the three major axes of the aircraft. All of these axes pass through the airplane's c.g.; therefore, all rotation is around the c.g..

- Maneuvering flight requires the use of all of the airplane's control surfaces in coordination with the power output of the powerplant to execute takeoffs, climbs, descents, turns, cruising flight, and landings.

BRING IT HOME

1. Describe in detail how a pilot's control movements in the cockpit lead to maneuvering the airplane in flight.
2. Discuss why the regulations restricting commercial airline passengers to their seats during takeoff and landing but allow them to move around during cruising flight are justifiable based on what you now know about how an airplane flies.

3. Create an aircraft design that produces maximum stability in flight.
4. Create an aircraft design that produces maximum maneuverability in flight.
5. Describe how the four forces of flight change during each major phase of a typical flight.

EXTRA MILE

1. Complete the Flight Simulator flight school lessons, and earn your Private Pilot's certificate.
2. Contact a local radio-controlled flying club, and attend a few meetings.

3. Take an introductory flight experience through the Experimental Aircraft Association's Young Eagles Program (www.eaa.org/youngeagles/) or a local flight school.

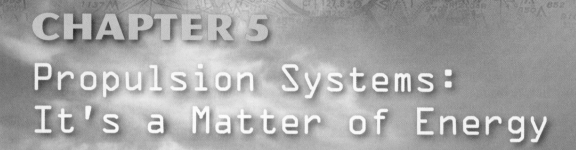

CHAPTER 5
Propulsion Systems:
It's a Matter of Energy

GPS DELUXE

| START LOCATION | DISTANCE | END LOCATION |

Menu

Before You Begin

Think about these questions as you consider the concepts in this chapter:

1 How can a glider fly without an engine?

2 Why doesn't flying faster always require a bigger engine?

3 What do propellers and wings have in common?

4 How does an internal combustion piston engine work?

5 Why are advanced materials required to produce a modern jet engine?

6 What are the differences among a turbojet, turbofan, and turboprop engine if they are all jets?

7 How does engine design change at extremely high speeds?

8 How does a rocket engine differ from an air-breathing engine?

The aerospace engineer's aircraft designs, like every other form of transportation, are constrained by the ability to provide a suitable power source for the vehicle. To fly at all, the vehicle requires the creation of a lifting force that can overcome the weight of the vehicle. To sustain flight, climb, or maneuver requires a source of energy that the engineer and pilot can use to change the position or motion of the craft. The evolution of our vehicle designs has always been limited and defined by our ability to design, build, test, and refine a propulsion system that transforms some type of energy into a mechanical force.

Energy is the ability to cause change, and it comes in many forms. The ability to cause change due to the motion of an object is obvious to anyone who has ever been hit by a thrown object or wiped out on a bicycle. Cooking clearly demonstrates the ability of heat to cause a change in the flavor and texture of our food. Who among us prefers the cake batter over the cake? Since the dawn of aviation, engineers have been tapping into the gravitational, kinetic, chemical, thermal, electrical, radiant, and nuclear forms of energy to power our aircraft and spacecraft. For this generation, aerospace engineers have designed, built, or are proposing vehicles that use every one of these seven forms as the source of energy for sustaining and maneuvering flight.

Although energy comes in many forms, we can simplify by describing any form of energy as either potential energy or kinetic energy. In this simplified view, potential energy is a stored energy that the vehicle can tap into in the future to sustain or maneuver in flight, whereas kinetic energy represents the energy in its current motion. These two are frequently exchanged one for the other. For example, almost all propulsion systems transform one form of energy, such as reducing the amount of stored chemical energy in the form of aviation fuel by burning it, to another form by lifting the entire vehicle to a higher altitude, which represents an increase in gravitational potential energy (Figure 5-1).

Figure 5-1 *MV-22 Osprey with its twin turboprop engines.*

Figure 5-2 *Artist's conception of a pulse rocket engine.*

It's also important to note that energy is always conserved. You can't make or destroy energy; you simply transform or transfer it from one object to another. However, this is always at a price. The transfer of energy always includes a loss to friction, drag, heat, or sound. Left to itself, this loss to an unwanted sink will continuously rob the aircraft of its combined energy of motion and altitude, bringing it firmly to the ground. Aerospace engineers design the propulsion system of the aircraft to be capable of providing the replacement energy for the aircraft to be able to continue doing what it's doing (Figure 5-2). Some aircraft gain the energy required from their surroundings while most bring the energy with them in form of a fuel. In this chapter, we focus on both the nature and the technology behind propulsion systems for aerospace vehicles.

FLYING WITHOUT AN ENGINE

Flying without power from an engine or motor is called gliding for an airplane and autorotation for rotorcraft like helicopters. At the start of a gliding flight, the aircraft has a limited and fixed amount of energy. This energy can be from motion, which is called kinetic, or elevated position known as gravitational potential. Once launched, the flight limits are defined by the tradeoff between these two forms.

Gliding flight can begin with a leap from an elevated site on a hill or mountain, release from another vehicle that has towed the vehicle to altitude, a rapid acceleration to high speed by a winch or bungee cord system on the ground, or if unlucky during a normal-powered flight that suffers an engine failure (Figure 5-3).

Figure 5-3 A towplane pulls a glider up to altitude.

© iStockphoto.com/Andy Gehrig

During the flight, the vehicle decreases its potential energy by descending through the air. To maintain airspeed, the vehicle has to descend at a rate that transforms the potential energy into kinetic energy at the same rate that energy is lost to the surrounding air. Although it's possible for the glider to descend more slowly than this sink rate, maintain altitude, or even climb, these all require that the energy lost to the surroundings, or gained in potential energy, must come from the vehicle's kinetic energy, and it slows down. By a similar argument, descending faster than the rate required for constant speed causes the vehicle to gain speed. Flight can be sustained indefinitely if the vehicle flies through a column or region of air that is rising faster than the vehicle's rate of descent through the air. Like an elevator, pilots have to be cautious because air currents can also flow toward the ground carrying the aircraft with them. To take advantage of these air currents, most modern "sport gliders" or sailplanes are equipped with a "rate of climb" instrument on the panel. A proficient sailplane pilot knows where to find thermals, ridge lifts, or mountain waves. All three of these provide dynamic energy to the glider in the form of wind currents created by solar or geographic features.

EFFICIENCY IN FLIGHT

Increasing the amount of flight time possible outside of regions of rising air demands that we either improve the aircraft's efficiency at producing lift without creating drag or that we run an engine to produce thrust that can overcome the drag in level flight. To improve efficiency, we try to minimize energy losses by creating very smooth, very streamlined aircraft with low drag coefficients so the maximum amount of energy remains to be applied to maintaining height and speed. To get the lowest drag coefficient with the greatest lift, most gliders employ a long and narrow high aspect ratio wing, which tends to be much lower in drag than low aspect ratio wings such as on an F-16. Let's take a moment to compare a performance glider to an F-16 jet fighter (Figures 5-4 and 5-5). Why do they look so different? Most variety in aircraft serves a purpose. An unpowered glider needs to stay in the air as long

Figure 5-4 *Glider with high aspect ratio wings.*

© iStockphoto.com/Aleksander Lorenz

Figure 5-5 *The Thunderbird's F-16 aircraft flying in formation.*

Courtesy of NASA

as possible on the mechanical energy at launch and what it can gain from the air, whereas an F-16 needs to be agile, maneuverable, and fast and carries a powerful jet engine and the chemical energy of jet fuel with it as it flies. Both aircraft could fly farther and "consume" less energy in the process with higher aspect ratio wings, but that is not the entire story.

It's a matter of the physics of **moments**: A short-winged F-16 can turn and bank rather quickly because the mass of the wings is close to the axis around which it rolls; however, the F-16's shorter wings demand that the aircraft maintain higher speeds to stay airborne. Most modern fighters can fly well in excess of Mach 1 (above the speed of sound). Try that with a glider! Although it's possible, the glider

Figure 5-6 *The definition of glide angle, glide ratio, and L/D ratio.*

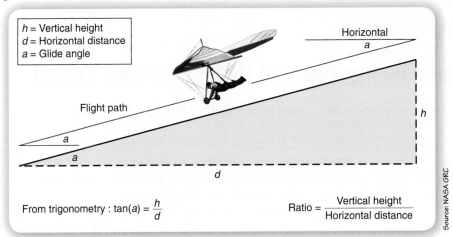

h = Vertical height
d = Horizontal distance
a = Glide angle

Horizontal
a

Flight path

h

a

a

d

From trigonometry : $\tan(a) = \dfrac{h}{d}$

$\text{Ratio} = \dfrac{\text{Vertical height}}{\text{Horizontal distance}}$

Source: NASA GRC

would have to lose altitude very rapidly to achieve these speeds, which isn't what a typical glider is designed to accomplish. Gliders maintain flight at much slower speeds. Longer span, shorter chord wings provide a lot less drag but are not very useful for sudden changes of direction or supersonic speeds.

Aspect ratio has a significant impact on the efficiency of an aircraft due to its effect on the lift to drag ratio (L/D), which affects an aircraft's capability to glide. The L/D ratio is equal to the glide ratio for the vehicle (Figure 5-6).

A glide ratio of 50:1 is not uncommon for modern sport gliders made from composite materials. What does this mean? The **glide ratio** number is proportional to how efficient the wing is at producing lift without creating drag. In the case of a glider with a 50:1 ratio, it means that the glider, in straight flight (not level!), will move 50 units of distance forward for every 1of the same unit of distance lost in altitude (50 feet forward for every 1 foot lost in altitude). Compare that to the Space Shuttle, which is the world's most inefficient glider with an L/D ratio of 3:1. No wonder they call it the flying brick. The unique mission of the Space Shuttle requires this glide ratio so that the pilot's can glide it from the high altitude and speeds of orbit to a flare for landing (Figure 5-7). The extremely low glide ratio indicates that to flare for landing, the vehicle has to produce a lot of drag, which is a good thing because this quickly slows the vehicle.

Figure 5-7 *The Space Shuttle's descent for landing is nonpowered.*

Courtesy of NASA

Engineering Moments: Human-Powered Flight: Gossamer Condor

During the 1970s, Dr. Paul McCready and friends designed and built the world's first successful human-powered aircraft capable of taking off, climbing above a 10 ft altitude, flying a 1.6 km long figure eight path, then once again climbing to above a 10 ft altitude to win the Kremer prize of almost $85,000.

To put this task into perspective, consider that the Wright Flyer had 12 HP to sustain a 750 lb loaded vehicle. This gives a power-to-weight ratio of 1 HP:62.5 lbs. Given that a human being can only sustain about ½ HP, this means that a vehicle as efficient as the Wright Flyer could only sustain flight if its total weight with a human engine aboard was less than 31.5 lbs! In practice, the final Gossamer Condor (Figure 5-8) weighed in at 70 lbs plus pilot Bryan Allen's 145 lbs for a total weight of 215 lbs. This gives a power-to-weight ratio of 1 HP:430 lbs or seven times more efficient than the Wright Flyer.

The connection to the Wright Flyer goes well beyond achieving a first in aviation. The Gossamer Condor used the same wing-warping technique as the Wright Flyer to produce most of its ability to turn.

McCready and Allen continued to develop their extremely lightweight human aircraft, eventually building the Gossamer Albatross, which successfully crossed the English Channel with a flight time of 2.8 hours. For more information, watch the 1978 best documentary film, *Flight of the Gossamer Condor*.

Figure 5-8 The Gossamer Condor hangs in the Air & Space Museum in Washington, DC.

© Cengage Learning 2012

PROPELLERS AND WINGS

Long before the first powered flight, windmills were used to generate energy for mills and pumps. For the early aviation inventors, it was logical to conclude that if the wind could push a blade or propeller, why not use the propeller to pull the wind and make their vehicles move? Mounting an engine on an airplane without a mechanism to transform its output into a mechanical push on the air is worthless. But combine a propeller or fan with the engine, and we have the **powerplant** for the vehicle. A powerplant is the combination of both a means to transform energy into mechanical form and a means to transfer it to the air.

Remember Newton's third law? If the airplane pushes on the air to make it move, then the air has to push on the airplane and can make it move! Also recall that Newton's second law tells us that $F = ma$, so to create a large force, we can move a large mass of air or change its speed by a large amount, or both. But the

Powerplant:

the combination of engine and propeller, turbines, or fans that transform energy from some source (chemical, electrical, mechanical) into thrust.

problem is that air is not very dense. So if the air pushing against the propeller has very little mass, it must experience a large change in speed to create enough thrust to push or pull the plane forward. This is the job of the powerplant of the aircraft. Comparing air to water makes this mass concept easier to understand. The propeller on a boat uses the same concept as the propeller on a plane; however, the boat propeller is much smaller and has its blades more angled to the path of rotation. Why is this? Again think back to Newton's third law. A boat propeller is pushing against water rather than air. The increased mass and density of water means a lot more "push" for the prop. But that's not the only factor; the propeller on a boat is more like a screw—a simple machine—which is quite different from the propeller on an airplane.

The airplane propeller has **pitch** *and* has the cross-sectional shapes of an airfoil just like the wing (Figure 5-9)! The spinning of a simple propeller creates a lift vector off each blade in the direction of flight. Because the tips of each blade are moving much faster through the air than at the root (remember that lift forces are related to the square of the velocity), a propeller blade with the same airfoil shape and size along its entire length produces much more force at the tip than at the root (Figure 5-10). This causes the blades to flex out of their plane of rotation and creates a strain that can snap the propeller. To produce a uniform thrust force everywhere along the span of the blade, the airfoil shape gets smaller and less pitched from root to tip.

Pitch:
the distance that a propeller or fan would advance during a single rotation if there was no slipping between it and the fluid through which it is rotating.

Figure 5-9 *Propeller blade profiles from root to tip.*
Source: FAA Pilot's Handbook of Aeronautical Knowledge

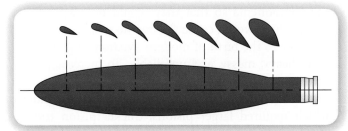

For both water and air propellers, the pitch on a prop determines the "bite" the propeller takes out of the air. A bigger bite—steeper pitch—requires more force from the engine, but the engine won't have to spin as much to maintain the force. A smaller bite—shallow pitch—won't put as much strain on the engine, but the engine will need to rotate the propeller a lot faster to maintain the force on the airplane.

Many modern planes have a variable pitch prop. The prop's pitch can be adjusted so that it is feathered—basically flat with no bite—or increased to take a larger bite. A variable pitch prop is useful for different phases of flight. During takeoff, we want a propeller that has less pitch so that the propeller spins more rapidly, affects the air more frequently, and is more gently accelerated. After we've reached cruising speed, we want a propeller that is more pitched so that we can slow down the engine to save fuel but maintain thrust. For fixed-pitched propellers, the engineers and mechanics have to choose a propeller that is a better climb prop or cruise prop; however, with a variable-pitch propeller, the aircraft has both modes available at all times.

Figure 5-10 *Relationship of travel distance and speed of various portions of a propeller blade.*

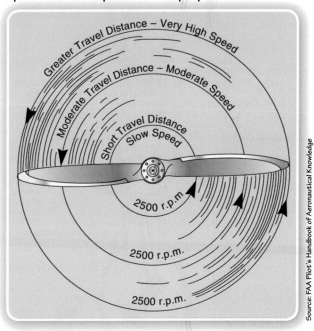

Source: FAA Pilot's Handbook of Aeronautical Knowledge

Figure 5-11 *A two-cylinder radially opposed engine from the early 1900s.*

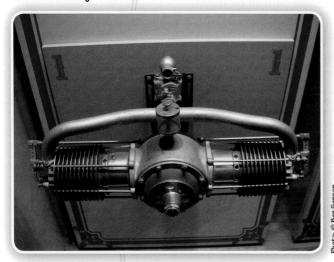

Photo: © Ben Senson

Figure 5-12 *Parts of a reciprocating internal combustion piston engine.*

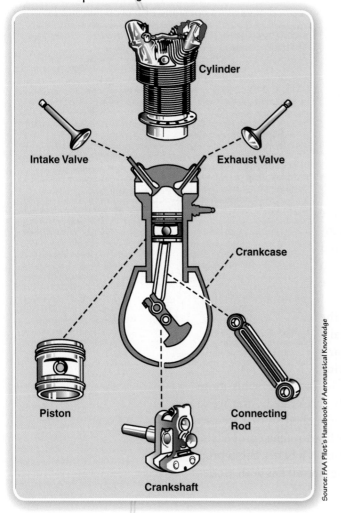

Source: FAA Pilot's Handbook of Aeronautical Knowledge

An engine is a machine that combines a chemical fuel, oxygen, and a heat source so that the energy released by burning the fuel is captured as the mechanical energy of motion (Figure 5-11). The earliest engines were very simple and poorly constructed. The first internal combustion engines had a movable piston that slid through a cylinder in a sequence that opened valves to intake the air and fuel mixture, then compressed it, and exposed it to a spark. As the gases burned, they expanded and pushed the piston downward in the cylinder and exhaust gases left the cylinder. This linear motion transformed into rotary motion by connecting the movable piston to a crankshaft with a piston rod. The crankshaft and cylinders were mounted in a crankcase or engine block to create an engine (Figure 5-12). The crankshaft was normally counterweighted so that as the piston moved in one direction through the cylinder, the vibration it caused was canceled out by the opposite motion of the crankshaft's mass. The engine also required a means of lubrication for all of the moving parts with oil and a way to cool the components so that they would not expand too much which would cause the engine to fail.

Prior to the birth of modern aviation, engines were loud, heavy, messy, and unreliable. To increase power, the designers mounted more than one cylinder to the same block and increased the volume of the cylinders, which required larger pistons, thicker piston rods, a sturdier crankshaft, and a much larger block. For aviation, this meant that at the turn of the century into the 1900s, an engine did not exist that was capable of powering an aircraft anywhere on planet Earth. In 1901, engines with sufficient power to drive the propellers were simply much too heavy.

In an aircraft, everything is related to weight—fuselage materials, wing coverings, wheels, and especially the engine. By 1903, the Wright brothers had to invent their own engine for their famous Kitty Hawk flight (Figure 5-13). The Wright engine represents the first time that light metal aluminum was used as a crankcase for an engine. Although the Wright engine was water cooled, the water evaporated as it was used and did not recirculate like today's water-cooled engines. The lightweight engine was extremely simple by today's standards yet produced an ample 12 HP to the Flyer's dual propellers at a weight of only about 180 lbs.

One of the most important benchmark measurements for aircraft engine performance is not simply

Figure 5-13 *A replica of the 1903 Wright Flyer engine.*

Photo: © Ben Senson

horsepower but the power-to-weight ratio. In other words, for each pound (or kilogram) of the engine, how much horsepower (or wattage) does it provide to the aircraft? A typical Toyota V-6 engine from a 1992 Camry has a weight of 469 lbs and delivers 185 HP.

Using these numbers, the Camry engine would have a power-to-weight ratio of 0.39 HP/lb. Compare this to the main thrusters of the Space Shuttle (6,700 lbs), which produce 375,000 HP (Figure 5-14)!

Piston Engines: Two Stroke Versus Four Stroke

Piston engines usually fall into two categories: two stroke or four stroke (Figures 5-15 and 5-16). The stroke defines the number of times the piston travels in one direction through the cylinder to complete one cycle of power output.

During two-stroke operation of a piston engine, the fuel and air mixture have to be brought into the combustion chamber (charging), then the mixture is compressed and ignited, which forces the piston down the cylinder and exhausts the resulting gases. During the first stroke, the piston moves downward in the cylinder to exhaust the hot gases from the last power stroke while the new fuel/air mixture is brought into the cylinder. During the second stroke, the piston moves upward in the cylinder and compresses and then ignites the fuel/air mixture. Timing of the ignition has to be carefully

Figure 5-14 **Main thrusters of the Space Shuttle.**

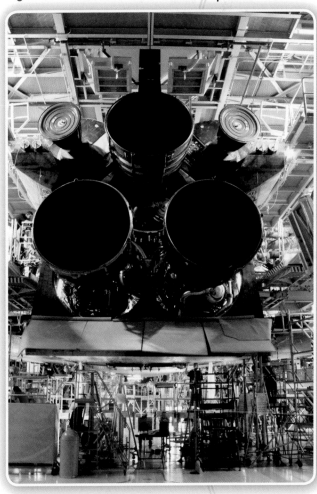

Courtesy of NASA

Figure 5-15 *Two-stroke engine cycles.*

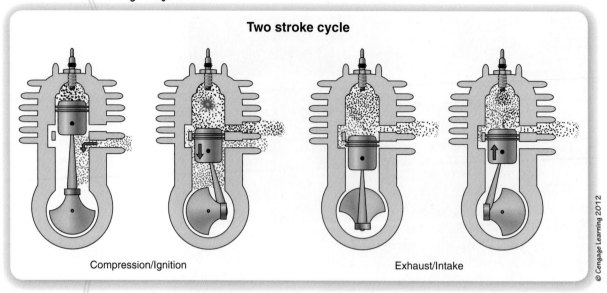

Two stroke cycle

Compression/Ignition Exhaust/Intake

© Cengage Learning 2012

Figure 5-16 *Four-stroke engine cycles.*

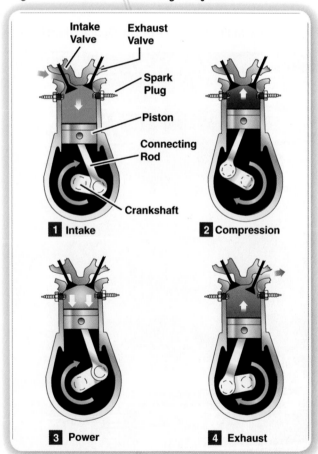

Intake Valve Exhaust Valve

Spark Plug

Piston

Connecting Rod

Crankshaft

1 Intake   **2** Compression

3 Power   **4** Exhaust

Source: FAA Pilot's Handbook of Aeronautical Knowledge

controlled so that it occurs when the piston is just starting to travel back downward in the cylinder.

A four-stroke engine requires the same processes of intake, compression, combustion, and exhaust; however, each is accomplished individually during a full stroke of the piston through the cylinder. The first stroke occurs as the piston moves downward in the cylinder, and the fuel/air mixture is pulled into the chamber through the intake valve. During the second stroke, the piston moves upward to compress the mixture after which a spark plug ignites the mixture. During the third stroke, the push from the expanding gases force the piston down through the cylinder. During the fourth stroke, the piston moves upward in the cylinder and pushes the hot gases out through the exhaust valve.

For both engines, single cylinder engines require the use of a flywheel to smooth out the energy output of the engine and sustain its rotation through the non-powered strokes until the next power stroke occurs. Multicylinder engines have the advantage of always having one of the cylinders in its power stroke, which reduces the weight of the rotating parts required to smooth out the engine's power output.

For this reason, the most common small engine for small general aviation aircraft is the four-cylinder, four-stroke, horizontally opposed, carbureted engine. Four-stroke engines also provide better fuel economy and reliability. Two-stroke engines are less complex and therefore tend to be lighter but do not have the efficiency of four-stroke engines and must rely on a fuel/oil mixture to help lubricate the engine.

Arranging the Pistons

Every engine arrangement has its advantages and disadvantages for aviation. Typically designs changed as the technology improved. Materials were stronger, lighter, and more heat tolerant, which helped improve the engine's power output, efficiency, smoothness of operation, and reliability.

Inline Air Cooled and Water Cooled

An inline engine mounted so that its crankshaft can directly turn a propeller has all of the cylinders mounted on the same side of the engine crankcase one behind the other (Figure 5-17). Interestingly, the first aviation engine was a horizontally mounted inline four-cylinder engine. This design worked typically well for an automobile engine but was difficult to adapt for aircraft engines. Trying to air cool an inline engine was very difficult as the nature of the piston arrangement blocked airflow around the three back cylinders, which usually left them overheating. Inline engines thus required water cooling, which added unnecessary weight to the aircraft. Design problems in cooling and weight of the inline led to the design of the rotary engine.

Figure 5-17 *A Curtiss Model K inline four-cylinder engine, 1911.*

Photo: © Ben Senson

Rotary

Engineers needed an inexpensive and lightweight alternative to the inline engines of early aviation. As aircraft saw use for military purposes the demands on the engine increased, requiring better performance for climbing and maneuverability and improved reliability. A rotary engine has pistons arranged radially and uniformly spaced around the crankshaft. The unusual feature is that the crankshaft is bolted to the body of the airplane, and the entire engine crankcase is spun around with the propeller (Figure 5-18). This design certainly solved the overheating problem, but created several others in the process. The large rotating mass of the engine created a gyroscopic effect on the aircraft, making it difficult to control in maneuvering flight. Accelerating the end of life for the rotary engine was one major flaw inherent

Figure 5-18 *A Murray Rotary 6 engine, 1912.*

Photo: © Ben Senson

in this type of design. As engineers tried to push the RPMs (revolutions per minute) to higher speeds, the increased rotation amplified the gyroscopic effects and increased centripetal effects on the engine, pooling oil in the cylinder heads or throwing it out of the engine entirely. Ironically, the design also meant that as the rotational speed increased, the drag caused by the rotating mass (initially to cool the engine) increased exponentially, reducing overall performance.

Radial

At first look, a radial engine seems similar to a rotary engine, but internally, they are very different (Figure 5-19). The radial engine's crankcase is mounted to the airframe, and the crankshaft and propeller are the only rotating objects. This solved the rotating mass problem of the rotary engines. Radial engines have an odd number of pistons with one piston being the master cylinder to which the other pistons are attached. This is a huge advantage for the radial because it was built around a small crankcase and cylinder heads that employed aluminum vanes to increase the area of cooling surfaces on each cylinder.

Figure 5-19 A Pratt and Whitney Hornet R-1860-B nine-cylinder radial engine, 1929.

Photo: © Ben Senson

Radial engines with their large frontal area are not very streamlined, but they do allow uniform cooling for all cylinders. Later designs incorporated stacks of radial cylinders on one engine. By slightly rotating each set of radial cylinders, all cylinders could get even cooling. One of the largest and most powerful radial engines ever built is the Lycoming XR-7755-3, which was designed for use on a bomber (Figure 5-20). It incorporated 4 sets of 9 radial cylinders connected to a common crankshaft and was turbocharged producing over 5000 HP for its 36 cylinders. The engine is the largest, most powerful reciprocating aircraft engine in the world; however, it was never flown on an airframe because the new gas turbine engines were much more reliable.

Radial engines solved the early problems of power to weight ratio and even cooling for all cylinders. The radial design, although shallow, did not have a very small frontal area. Cowlings helped reduce the drag effect somewhat, but as more aircraft designers turned to sleeker and more aerodynamic fuselages for fighter aircraft, an alternative was needed.

Figure 5-20 A 5000 HP Lycoming XR-7755-3 radial engine with 36 cylinders, 1946.

Photo: © Ben Senson

V-Type Engines

The first patent for a V-piston engine—in which pairs of pistons are arranged to form a "V" shape—appears before the end of the 19th century. It didn't take long for its first use in aircraft just three years after the Wright brothers' first powered flight. Mainstream use of the V-engines was never fully realized because early designs tended to be heavy with marginal horsepower. Later technology improved the horsepower to weight ratio, and V-type engines saw increased use because streamlined airplane designs and faster speeds required smaller frontal area engines (Figure 5-21). The Liberty 12-cylinder, air-cooled engine, designed during

Figure 5-21 The 1490 HP Packard Merlin V-1650-7 placed its 12 cylinders into a V configuration, 1933.

Photo: © Ben Senson

World War I, could produce 400 HP. It was one of the most powerful aircraft engines for the time but still had an enormous weight. The Liberty 12 was used in the DeHavilland DH-4, which was a British built but U.S. redesigned aircraft that was used primarily for observation and day bombing. Through the years following the Great War, the V-type engines were mainly used for sleek racing type aircraft until World War II. During WWII, almost all fighter aircraft employed a V-type engine design.

Horizontally Opposed

There are several variations on the horizontally opposed engine. All flat engines have the pistons directly across from each other usually with 180° angles (Figure 5-22). Where the variety comes into play is how the pistons are attached to the crankshaft. If pairs of pistons share a common pin on the crankshaft, then the pistons will move in opposite cycles. This is known as a 180 V-Engine. The "boxer" type of flat engine has pistons that reach compression at the same time. In other words, the compression cycles are at the same phases across pairs of pistons, and each pair of pistons will either be fully extended or contracted at the same time.

Figure 5-22 A Lycoming four-cylinder, horizontally opposed, normally aspirated engine.

Photo: © Ben Senson

Horizontally opposed engines are common in smaller aircraft for several reasons. First, because they are flatter in design, they have better aerodynamic characteristics, and are preferred over the larger radial engines. Flat engines have good natural balance because pistons counteract the opposite piston within the same plane of motion—no need for heavy counterweights on the crankshaft. Air-cooling is easily accomplished on flat engines because the pistons have plenty of room around them.

Experimental Piston Engines

Piston engine design for small aircraft has not changed much over the past several decades. Aircraft engine designs must be certified and thoroughly tested before approval for flight, which can be costly and time consuming. In 2005, there were only

224,352 registered general aviation aircraft. That number includes both piston engine and private jets. Compare that to the number of registered automobiles for the same year, which was more than 247 million. In general, this makes the market for a new airplane engine design not economically feasible. Because of this, engine designs that were drafted in the 1950s and 1960s are still used today with only minor modifications to the overall design.

Pilots and designers have continued over the years to modify existing designs or create their own designs for aircraft and engines. These aircraft are assigned to the Experimental Aircraft category, which has allowed aviators options for engine use. Several automotive engines have been adopted for aviation use in this category without going through the aviation certification process. The air-cooled VW flat engines of the original beetle era regularly see adaptation to home-built aircraft (Figure 5-23).

Although three Wankel type engines saw testing and use on small aircraft, the full employment as a certified aircraft engine did not happen. But the Wankel continues to see adaptation to aviation use from home builders due to its relative simplicity compared to reciprocating piston engines, reliability, light weight, and fuel economy. Most of these adaptations spring from the Mazda version of the Wankel engine.

Figure 5-23 *The AeroVee Conversion engine kit uses a VW block with many kit components engineered to transform it into an aviation engine.*

Courtesy of Sonex Aircraft, LLC

Increasing Power at Altitude

Throughout the development of the reciprocating engine, aerospace engineers have been challenged by the limitation represented by the atmosphere. As an aircraft climbs higher into the atmosphere, the air gets thinner and thinner. This means that to run efficiently, the engine has to burn less and less fuel, reducing power output of the engine, or more air has to be fed into the engine. To accomplish this second goal, engineers developed the **turbocharger** (Figure 5-24).

Turbocharger:
a device that compresses the air entering an engine so that it can combust more fuel per unit time, thus increasing the power output of the engine.

Figure 5-24 *The components in a turbocharging system.*

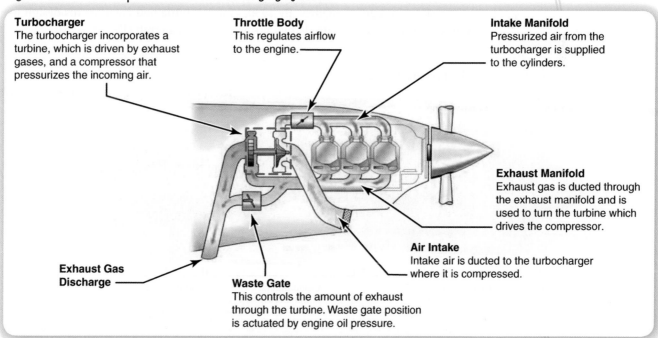

Turbocharger
The turbocharger incorporates a turbine, which is driven by exhaust gases, and a compressor that pressurizes the incoming air.

Throttle Body
This regulates airflow to the engine.

Intake Manifold
Pressurized air from the turbocharger is supplied to the cylinders.

Exhaust Manifold
Exhaust gas is ducted through the exhaust manifold and is used to turn the turbine which drives the compressor.

Air Intake
Intake air is ducted to the turbocharger where it is compressed.

Exhaust Gas Discharge

Waste Gate
This controls the amount of exhaust through the turbine. Waste gate position is actuated by engine oil pressure.

Source: FAA Pilot's Handbook of Aeronautical Knowledge

Figure 5-25 *Power output of a normally aspirated engine compared to a single-stage, two-speed supercharged engine.*

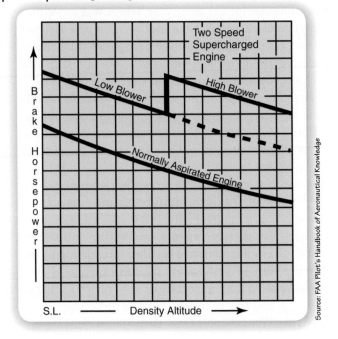

A turbocharger is a device that captures some of the energy from the exhausted gases leaving the engine to rotate a compressor in the intake path of the air for combustion. By compressing the air before it enters the engine, the aerospace engineer can make the engine run as if it was flying at a lower altitude (Figure 5-25).

MATERIALS SCIENCE AND THE JET TURBINE ENGINE

More commonly known as a jet engine, early concepts of the gas turbine engine design by F. Stolze first appeared in 1872. His design never worked under its own power, but the fundamental concepts of the engine operation were later refined and used in successful commercial engines. By 1930, Frank Whittle had patented his design for the first gas turbine to be used as jet propulsion in an aircraft (Figure 5-26).

Figure 5-26 *Basic components of a turbine engine.*

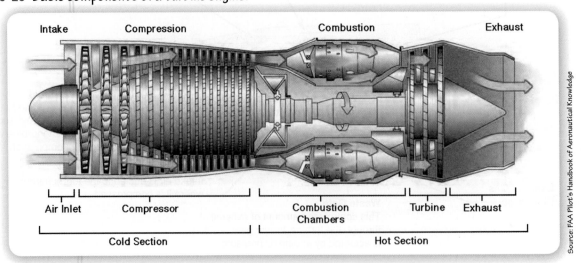

All gas turbine engines—axial or centrifugal—follow the same processes for energy output. Incoming air is compressed using a series of alternating rotating turbine blades and stationary stator vane stages that accelerate and compress the air. The compressed air is forced into the combustion area where a fuel is mixed with the compressed air and ignited, causing an expansion of the exhaust gases. This expanding gas accelerates past another set of turbine and stator vanes, which are coupled by a shaft forward to drive the compressor stages. Depending on the amount of fuel added to the system, this can maintain the compressor speed or accelerate it, creating more power. In turbojets, the accelerated hot air is compressed through a nozzle, which increases thrust. In other turbine applications, the main shaft can be connected to a large fan (turbofan) or to an axle and gear reduction box to drive a much slower propeller or rotor (turboprop or turboshaft for helicopters).

> **Nozzle:**
> a shaped channel that is designed to control and guide the exhaust gases from an engine.

Following are the advantages of gas turbine engines:

▶ Continuous combustion process that allows 20 times as much fuel to be combusted for an equivalent combustion chamber volume and time period

▶ Very high power-to-weight ratio, compared to reciprocating engines

▶ Increased horsepower with increased RPM

▶ Smaller than most reciprocating engines of the same power rating

▶ Moves in one direction only, with far less vibration than a reciprocating engine

▶ Fewer moving parts than reciprocating engines

▶ Low lubricating oil cost and consumption

Following are the disadvantages of gas turbine engines:

▶ High material cost to handle the extreme temperatures and rates of rotation

▶ Much less efficient at lower RPMs

▶ Response lag

▶ Highly susceptible to ingestion of FOD (foreign object debris) such as dust, dirt, and scrap particles

TURBOJET, TURBOFAN, AND TURBOPROP ENGINES: MATCHING THE ENVELOPE

Aircraft operate over a wide range of altitudes and speeds. For any specific aircraft, there is a limit to the combined altitude and speed that the aircraft is capable of reaching. This flight envelope has to be matched to the propulsion system that is used on the aircraft (Figure 5-27).

The three primary designs for engines that use a gas turbine or jet engine core as their primary power source are the turbojet, the turboprop, and the turbofan. For a turbojet, the exhaust gases of the engine directly produce the thrust to accelerate the aircraft forward. This is very effective at high speeds because the exhaust gases leave the engine at very high speeds. To achieve even greater thrust and thus higher speeds, a turbojet engine can have an afterburner, or reheat unit, installed that adds additional fuel to the exhaust of the engine after the combustion chamber. This boosts the exhaust temperature and speed to provide additional thrust.

Figure 5-27 The performance characteristics of various air breathing engines.

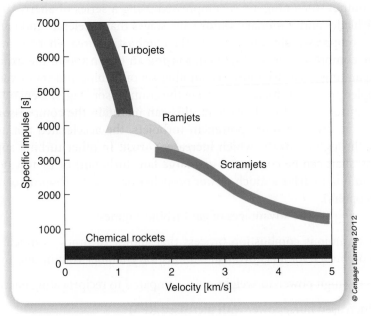

However, if the aircraft is going to operate at subsonic speeds for its entire flight, then it is far more efficient to move a larger volume of air at a slower speed. To meet this requirement for most commercial aviation, the aerospace engineer would use a turbofan. In the modern high-bypass **turbofan** engine, six times as much air bypasses the engine core and produces thrust without being involved with the combustion of fuel. At even slower speeds, a fan moves the air too rapidly to efficiently convert the gas turbine's power into thrust. At speeds of a few hundred miles per hour (or knots), the turbine core's power output is run through a gear box to slow down its rotation and then to a propeller to form a **turboprop** engine (Figure 5-28).

Figure 5-28 The four common types of gas turbine engines

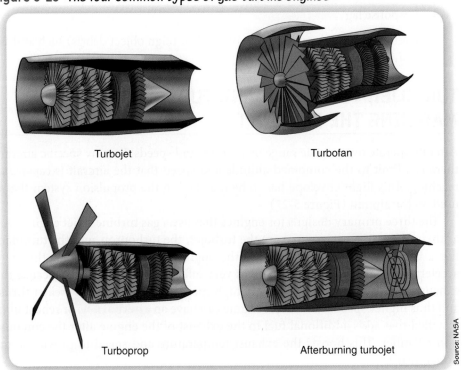

HIGH-SPEED ENGINES: RAMJETS AND SCRAMJETS

At extremely high speeds that are more than twice the speed of sound, we require a different type of engine. This realm of hypersonic flight requires a ramjet or scramjet engine (Figure 5-29). Ramjets and scramjets are similar to gas turbine engines in that they are continuous combustion engines. After that, the similarities end. Ramjets and scramjets usually have no internal moving parts because the engines use the extremely rapid movement of the engine through the air and the shape of the engine inlet to cause the air to compress prior to entering the combustion chamber. Because of this, both engines require speeds in excess of Mach 1, or the speed of sound. Ramjets typically operate at speeds between Mach 3 and Mach 5, whereas scramjets can operate from Mach 5 all the way to orbital speeds of greater than Mach 24 (Figure 5-30). In a **ramjet**, the velocity of the air decreases from the supersonic speed of the freestream flow to subsonic, or slower than the speed of sound, in the burning chamber. Even if the aircraft is traveling at supersonic speeds, the design of the inlet and diffuser of a ramjet helps to slow down and guide the supersonic air into the engine. **Scramjet** velocities in the

Figure 5-29 Ramjet and scramjet engines work at supersonic speeds; however, the airflow through a ramjet engine slows to a speed below the speed of sound as it passes through the engine while airflow through the scramjet remains supersonic.

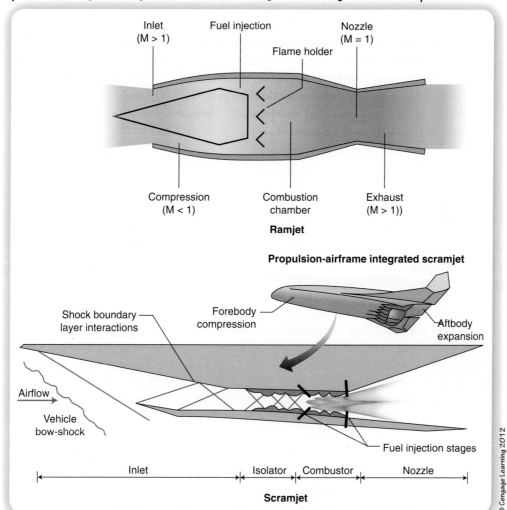

© Cengage Learning 2012

Figure 5-30 *The second X-43A hypersonic research vehicle, mounted under the right wing of the B-52B launch aircraft, uses a scramjet engine to reach speeds in excess of Mach 7.*

Courtesy of NASA Dryden Space Flight Center

chamber are supersonic, faster than the speed of sound (Supersonic Combustion ramjet). Both varieties of ramjets must be started at high velocities. The scramjet holds the current speed record for an air-breathing engine at Mach 10 (about 11,000 km/hr). So far, the best method to gain the high speed needed for compression is by rocket-assisted acceleration. However, gas cannon and rail gun launch systems are in development.

Point of Interest

The Pulse Jet and Ramjet

Ramjets and pulsejets are often confused, but they are actually quite different types of propulsion. A pulsejet relies on a valve system to mix and ignite the air/fuel mix before combustion and subsequent ejection—all as a series of pulses; a ramjet is a very simple device that literally consists of an intake, combustion, and nozzle system in which the cycle pressure rise is achieved purely by ram compression. Consequently, a separate propulsion system is needed to accelerate the vehicle to speeds at which the ramjet can take over. A ramjet is literally just a tube into which fuel is continually fed and ignited. The ramjet does not use a pulsed mode of combustion, but rather continuous combustion.

Ramjets and scramjets are mechanically the simplest of all aviation propulsion. But they are the most complex aerodynamically. There are no turbines or valves to control combustion pressure. The incoming gases and compression must be maintained by the forward speed of the engine and compressibility of the incoming air. Careful study of the inlet airflow and aerodynamics of the diffuser cone is critical for a successful ramjet.

Thrust with ramjets occurs because the exiting gas flow is at a higher velocity than the incoming air. This is accomplished by burning a small amount of fuel and using a nozzle to constrict the exiting gas, thus increasing its velocity. A scramjet still has a rear nozzle to speed up the exiting gas even though it's already traveling at supersonic speeds as the thrust of the engine is determined by how much faster the exhaust is leaving the engine compared to the aircraft's overall speed.

ROCKET ENGINES

The main advantage of **rocket engines** is their ability to work in the vacuum of high altitudes and outer space. In other words, a rocket engine carries its own fuel and oxidizer. Recall from the beginning of this section that combustion in an engine requires fuel and oxygen. A combustion chamber allows the hot gases to mix and ignite. These hot gases escape the combustion chamber and are forced through a narrow nozzle but expand on the other end. As we know from Bernoulli, this accelerates the speed of the hot gases escaping, and the opposite reactions propel the rocket forward. Rockets, whether solid based or liquid based, carry their own oxygen to burn, which is why they are the top choice for sending objects into the vacuum of space (Figure 5-31).

Rocket engine:
a reaction engine in which the original thrust force is created by combusting and exhausting fuel and oxidizer, which are entirely derived from materials carried on board the vehicle.

Figure 5-31 *The Space Shuttle Atlantis rises majestically on the thrust of its combined solid rocket boosters and liquid-fueled main thrusters to start mission STS-122 to the International Space Station.*

Courtesy of NASA

Solid Fuel

A **solid fuel** rocket motor has fuel and oxidizer mixed into a solid propellant that is packed into the casing of the motor. The combustion chamber is initially formed by a cavity hollowed out of a portion of the solid material and can be round, square, star-shaped, or more complexly formed depending on how much surface area is required to support the rate of combustion required. The faster the fuel burns, the greater the thrust that is produced. As the solid rocket fuel burns along the inside of the rocket body, the entire tube becomes the combustion chamber. Solid-fueled rockets still have nozzles to increase the velocity of the escaping exhaust and to guide it in the direction opposite of the thrust needed.

Once ignited, a solid fuel rocket motor cannot be throttled or turned off because it has its own oxidizer mixed with the fuel. The dangers of this type of rocket were evident in the Space Shuttle Challenger disaster, in which the hot gas eroded through an o-ring seal and quickly burned into the liquid rocket engines external fuel tank, causing the vehicle to fail.

Solid fuel rocket motors are used when the launch has to occur at a moment's notice or when the small size and portability of the rocket is an important part of accomplishing its mission.

Point of Interest

Engineering Moment: Learning from Disaster—O-rings and Challenger

In 1986, the Space Shuttle Challenger was launched from Kennedy Space Flight Center but never reached orbit. The launch had been put off by prior mission delays, bad weather at its emergency landing sites, and hardware issues on a hatch. By the scheduled time of launch at 11:38 AM, the temperature was below the designed operational temperature for the o-ring seals that are used to prevent the escape of the combustion gases from between the joints of the segments that make up a solid rocket booster (SRB). At the time of launch, both the primary and secondary o-rings failed, allowing a jet of rocket exhaust to shoot directly onto the external tank that contains millions of TNT pounds equivalent of explosive fuel and oxygen.

The external tank exploded, the shuttle shifted off its flight path, and aerodynamic forces destroyed the craft with all crew onboard lost. The accident would have been prevented by launching only if conditions were within the constraints of the engineer's original designed performance capabilities.

Liquid Fuel

Liquid fuel rockets are relatively new compared to solid fuel rockets. As the name implies, the fuel and oxidizer are in liquid form. The liquids are pumped by turbines to be mixed together and ignited in the combustion chamber. The combustion chamber of a liquid rocket does not change its shape throughout the burn as occurs with a solid fuel rocket. The actual size of a liquid fuel rocket combustion chamber can be relatively small; however, the fuel storage takes up considerable space. The fuel of choice for liquid rockets is hydrogen with oxygen as the oxidizer. To keep hydrogen in a liquid state it must be pressurized and kept at a chilly $-423.17\ ^\circ\text{F}$ ($-252.87\ ^\circ\text{C}$). Oxygen can stay liquid at much warmer temperatures, $-297.3\ ^\circ\text{F}$ ($-183.0\ ^\circ\text{C}$). Compare this to a typical home freezer that can only maintain a temperature of $0\ ^\circ\text{F}$ ($-18\ ^\circ\text{C}$).

Liquid rocket engines are mechanically complex due to the need for the rapid pumping and proper mixing of the liquids and the extreme temperatures needed to store the liquids. Liquid rockets hold a huge advantage over solid rockets because the combustion can be controlled or even shut off. By throttling the amount of liquids that come into the chamber, a liquid engine can be controlled much in the same way you control the speed of your car. Liquid rockets are used where the timing of the launch can be decided in advance, when the advantages of a higher power-to-weight ratio is necessary, or when the controllable burn is essential to accomplishing the mission.

Point of Interest

Engineering Moments: Chasing the X-Prize with Hybrid Fuel Rockets

In the race to develop the world's first nongovernmental spaceship, Burt Rutan and his Scaled Composites crew invented their own hybrid rocket engine (Figure 5-32). Part solid, part liquid rocket, their engine uses hydroxyl-terminated polybutadiene (tire rubber) and nitrous oxide (laughing gas) as fuel and oxidizer, respectively.

The main advantages of the system are that the solid fuel is stable and safe to store while the liquid oxidizer vaporizes and self-pressurizes at environmental temperatures so that no complicated pumps are required to deliver it to the combustion chamber.

On October 4, 2004, SpaceShipOne became the winner of the X-prize as the world's first commercial spacecraft to leave Earth's atmosphere twice within a 14-day period with 3 people on board. Today, the story continues with the development of the world's first space line—Virgin Galactic. See them at www.virgingalactic.com/

Figure 5-32 The hybrid engine from Spaceship One, the first nongovernmental spacecraft to reach outer space twice within a two-week time period to win the Ansari X-Prize in 2004.

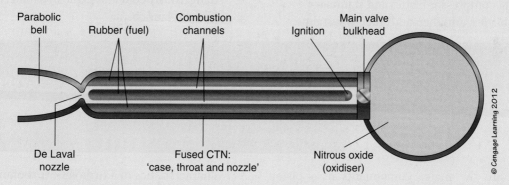

© Cengage Learning 2012

SUMMARY

- Early aviation engines suffered from low power-to-weight ratios, inadequate cooling, and unreliable operation.

- The internal combustion engine has many variations, but the three primary versions are piston engines, gas turbine (jet) engines, and rockets.

- The four-stroke engine is the most common type of piston engine due to its smooth, reliable, efficient operation.

- The four primary strokes of a piston engine are the intact, compression, combustion (power), and exhaust strokes.

- Turbochargers compress air flowing into a reciprocating engine to support burning fuel more quickly and increase the power output of the engine at any given altitude.

- The gas turbine jet engine is a continuous combustion engine that compresses air with rotating compressor blades and stationary stator blades, and then adds fuel to sustain

combustion. Some of the exhaust gas energy is captured with rotating turbine and stationary stator vanes to obtain the energy required to spin the compressor stages while the remainder accelerates through the nozzle to produce thrust.

- The efficiency of a gas turbine engine can be improved by accelerating additional air with devices driven by the engine's shaft such as a cowled fan for high-speed turbofan engines or a propeller for slower turboprop aircraft.

- Rocket engines carry their own oxygen supply with them so they can operate at high altitudes and in the vacuum of outer space.

- Solid fuel rocket motors have the advantage that they can be ignited at a moment's notice but the disadvantage that they cannot be throttled or extinguished until combustion is complete.

- Liquid fuel rockets have the advantage of controllable burns but require handling dangerously cold and explosive materials just prior to launch.

BRING IT HOME

1. What are the major components of a piston engine?
2. What is the difference between a rotary and a radial engine?
3. Describe what is happening during each of the four strokes of a common internal combustion engine.
4. Complete Internet research to describe the primary differences between the design and flight characteristics of a turbojet-, turbofan-, and turboprop-powered aircraft.
5. Research the types of jet and rocket engine nozzle shapes and why each is used.
6. Construct a commercially available rocket from a kit, and then launch it. Be certain to follow all the safety guidelines of the National Academy of Rocketry (www.nar.org/safety.html).

EXTRA MILE

1. "Necessity is the mother of invention." Describe the history of aviation engine development in terms of the needs that drove them to higher performance levels and greater reliability.

2. Design a custom rocket using software such as NASA's Rocket Modeler or Apogee's RockSim software (www.apogeerockets.com/rocksim.asp). A free, 30-day trial edition is available. After the design has been tested as stable, construct the rocket out of commercially available components, and fly your creation. Be certain to follow all the safety guidelines of the National Academy of Rocketry (www.nar.org/safety.html)

3. Explore the role that the Experimental Aircraft Association has played in supporting the development of aerospace engines and aircraft.

4. Research the development of human exploration of space during the Mercury, Gemini, Apollo, Space Shuttle, International Space Station, and Constellation program eras.

CHAPTER 6
Avionics and Flight Systems

GPS DELUXE

Menu	START LOCATION	DISTANCE	END LOCATION

Before You Begin

Think about these questions as you consider the concepts in this chapter:

1 What is pilotage and how is it used to navigate from one location to another?

2 How is dead reckoning an improvement over pilotage as a navigational aid?

3 What is the limiting factor for accurately navigating with pilotage and dead reckoning?

4 How were radio signals used first to create navigation aids for pilots?

5 What does VOR stand for, and how do they operate to create narrow radio airways in the sky?

6 Why would an airport install an instrument landing system on some of their runways?

7 What role does radar play in helping pilots navigate during routine flights?

8 What is "global" about the Global Positioning System, and how is this coverage achieved?

9 How is a synthetic vision system different from a traditional instrument panel?

The history of aviation is a tale of pilots and aircraft getting lost less and less frequently. By the late 1920s, aircraft had become reliable enough to take off at dawn and fly throughout the day. But as night began to fall, the airplanes landed, and their passengers transferred to trains for overnight travel. To become competitive with rail transportation and shipping, the early aircraft had to climb into the featureless atmosphere for overnight flights and flight in weather that was inhospitable. And to make nighttime and all-weather air travel possible, engineers had to develop techniques for tracking aircraft location.

Accuracy in navigation dramatically increases the effective range over which an aircraft can operate, improves safety by allowing aircraft to find their destination or divert to the nearest and most suitable airport, and increases efficiency by decreasing fuel consumption and time aloft. As human innovation and technology has dramatically altered every other arena of our lives, so too have they changed how we find our way around the world. Aviation has evolved from the visual art of nap-of-the-Earth flying to routine flight operations based almost entirely on our ability to extend our senses through radio-, radar-, and satellite-based systems.

And yet through the entire evolution, aviation has held tightly to the concept of redundancy and safety so that when the newfangled device has a glitch, the pilot in command has both the technology and the skills required to safely operate the aircraft.

Boeing 747-400 simulator cockpit.

Courtesy of NASA/Ames Research Center

THE COMPASS, GYROSCOPES, AND PRESSURE GAUGES IN THE COCKPIT

The simplest navigational system is based on human eyes: Take off . . . follow landmarks . . . hope you do not lose sight of the ground. However, it can be very difficult to see an airport from the air, especially if you aren't close to the airport, and you don't know which way to look. Three technologies provided the basis for our first steps into the void beyond the visual and into instrument flight conditions. The use of the ancient mariner's compass and the addition of gyroscopically stabilized and pressure sensing instrumentation allowed us to achieve the goal of an on-time departure at night and into most weather conditions.

Pilotage

When you take off from almost any airport, as soon as you are airborne, you can see quite far away. If you pick a feature somewhere out in the distance, fly toward it, and when you get there repeat the process, you are navigating by simply looking for ground features, which is called pilotage. When you are flying close to home, this works quite well because you are familiar with the major highways, rivers, lakes, and cities where you live, but what happens when you actually want to go somewhere far away or where you have never been before. As shown in Figure 6-1, pilotage is based on using charts and maps to select a sufficient number of visual reference points, features on the ground, or waypoints that can guide the pilot from one place to another in connect-the-dot fashion. The technique is simple but very limited because pilots burned a lot of fuel zigzagging across the countryside and were forced to land as soon as they couldn't see the ground clearly. Fog, haze, rain, snow, or simple darkness prevented effective navigation by pilotage.

Figure 6-1 Pilotage and dead reckoning navigational techniques.

Dead Reckoning

Fly due east at 50 mph for an hour and you ought to be 50 miles east of where you started. This is the logic underlying navigation by dead reckoning. The key to dead reckoning is the relationship between the aircraft's speed and heading and the time spent in flight. Combined with accurate maps, the technique allows pilots to fly without visual reference to the ground over very long distances. However, the technique is limited to the accuracy with which the pilot can maintain heading and airspeed. This is why students seeking Private Pilot certification are required to demonstrate the ability to maintain an altitude, heading, and an airspeed during their flight proficiency test. Any inaccuracies accumulate throughout the flight and add up to a larger error in navigation. Thus, all flights end with a landing under visual flight rules that require the pilot to visually identify the runway or other approach aids prior to completing the landing.

Figure 6-2 The main flight instruments on the instrument panel are frequently referred to as the "Six Pack".

6-Pack Plus One

To fly an airplane under instrument flight rules requires the aircraft to have a radio for communication with air traffic control and seven critical instruments that help the pilot maintain control over its attitude and flight path (see Figure 6-2). The instruments are generally clustered on the panel and look like a six-pack of soft drinks viewed from above. Three of the instruments use gyroscopes, while the other three instruments measure and compare air pressure to drive their indications. The last required instrument is a clock that indicates time to the second (see Figure 6-3 for labels).

Figure 6-3 The primary flight instruments required for flights into instrument conditions. A clock with seconds indicated is also required.

Gyroscopes operate on the principle of conservation of angular momentum. A massive object that is rotating in space will maintain its orientation. This is referred to as gyroscopic stabilization. If external forces are applied to the gyroscope, it experiences a torque that causes it to precess (see Figure 6-4). Precession is the wobbling of a top, an obvious change in orientation. To be useful for navigation, we make the gyroscope massive, rotate it very rapidly, and mount it in a nearly frictionless bracket. If we can minimize, or correct, for precession, a gyroscope provides a reference for orientation.

Figure 6-4 Gyroscopes hold their orientation in space when mounted on low-friction gimbals but move in precession whenever a force is applied perpendicular to the axis of rotation.

Source: Pilot's Handbook of Aeronautical Knowledge, FAA-H 8083-25

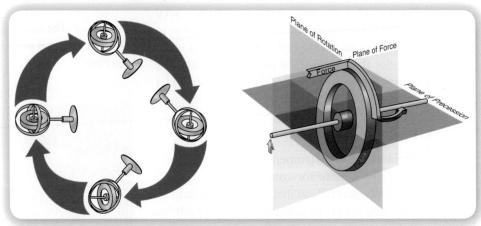

Figure 6-5 The parts of the Pitot-Static system.

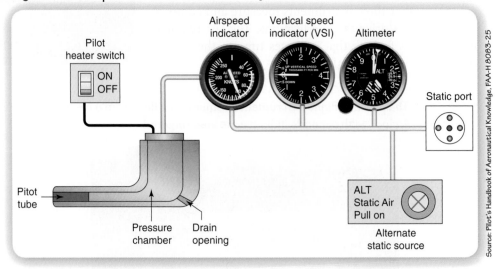

Source: Pilot's Handbook of Aeronautical Knowledge. FAA-H 8083-25

Airspeed Indicator:

a Pitot-Static instrument that measures and communicates to the pilot the relative speed of the air and the vehicle. Note that this is commonly different from the ground speed of the vehicle due to wind.

Figure 6-6 The airspeed indicator measures the velocity of the relative wind.

© Image by X-Plane

Pressure-based instruments use the push of the air to directly move mechanisms that eventually turn a dial on the instrument face. Shown in Figure 6-5, this is called the Pitot-Static system because the instruments monitor the pressure of both the moving (Pitot or ram) air and the static pressure of the stagnant air. Today, aircraft roll off the assembly line with modern flat panel instrumentation that uses electronic sensors for the pressure sensing.

Airspeed

Air molecules push against everything that they contact. The random motion of the air molecules allows them to create pressure even when the overall mass of the air is not in motion. This pressure is called static pressure. If the air is also in motion, it produces more pressure due to its motion as a group. This pressure is dynamic, ram, or Pitot pressure.

Inside the **Airspeed Indicator** is a pressure chamber or cell (see Figure 6-6). The static side of the cell is the instrument housing itself, which is connected by a tube to a place where the outside air is flowing perpendicular to the opening

Figure 6-7 The ram air pressure from the Pitot tube is compared to the static pressure to determine airspeed.

Source: Pilot's Handbook of Aeronautical Knowledge, FAA-H 8083-25

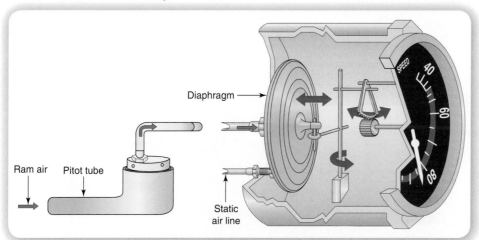

so that its motion isn't measured. The inside of the cell is connected by a tube to a Pitot tube that is mounted in a location undisturbed by prop wash. The Pitot tube points directly into the air so that it measures both the static and dynamic pressure (see Figure 6-7).

The static pressure exists on both sides of the membrane and cancels out leaving only the Pitot dynamic pressure on the inside of the cell. The faster the air flows into the Pitot tube, the higher the dynamic pressure and the more the cell's walls are pushed outward. The expansion of the cell causes the needle on the instrument to deflect giving us a reading for the indicated air speed (IAS).

Altitude

The **altimeter** contains an aneroid, a sealed cell with air trapped inside (see Figure 6-8). The aneroid itself is squeezed so that it is thinner when the static pressure around it is higher and thicker if the static pressure is lower (see Figure 6-9). Because air pressure decreases with altitude, the aneroid's expansion can be turned into a dial indication for altitude. Air pressure can also change with temperature (see Figure 6-10).

You may have noticed that the weather map has large Hs and Ls on it to indicate what is happening with local air pressure (see Figure 6-11). Because the pressure at Earth's surface varies from day to day, the altimeter has a knob that allows for calibrating to the current weather conditions. This correction is reported to pilots

GRASS STRIP ADVENTURE

Investigate the Pitot-Static system at *www.luizmonteiro.com/ Learning_Pitot_Sim.aspx*.

Figure 6-8 The altimeter measures the elevation of the aircraft compared to mean sea level (MSL) if it is adjusted to reflect local atmosphere pressure at the surface.

© Image by X-Plane

Figure 6-9 The altimeter contains a sealed aneroid (chamber with a calibrated pressure) to which static pressure is compared. Because static pressure decreases with altitude, the instrument indicates a higher altitude as static pressure decreases.

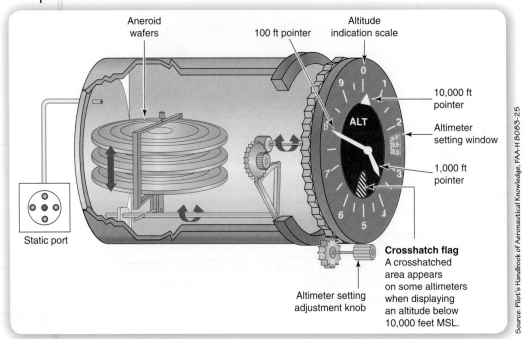

Aneroid wafers

100 ft pointer

Altitude indication scale

10,000 ft pointer

Altimeter setting window

1,000 ft pointer

Crosshatch flag
A crosshatched area appears on some altimeters when displaying an altitude below 10,000 feet MSL.

Static port

Altimeter setting adjustment knob

Source: Pilot's Handbook of Aeronautical Knowledge, FAA-H 8083-25

Figure 6-10 Pilots must update the altimeter's reference pressure setting frequently to avoid flying the constant pressure slope downward toward the surface.

Source: From AHRENS. *Meteorology Today: An Introduction to Weather, Climate, and the Environment,* 9e. © 2008 Brooks Cole, a part of Cengage Learning, Inc. Reproduced with permission. www.cengage.com/permissions

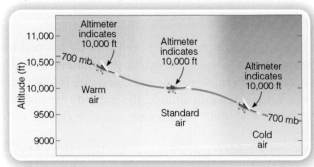

Figure 6-11 In the absence of temperature changes, pressure surfaces can still dip toward Earth as the aircraft flies from high- to low-pressure areas. This leads to the pilot's reminder, "Flying high to low, look out below!"

Source: From AHRENS. *Meteorology Today: An Introduction to Weather, Climate, and the Environment,* 9e. © 2008 Brooks Cole, a part of Cengage Learning, Inc. Reproduced with permission. www.cengage.com/permissions

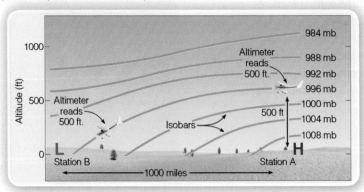

via an Automated Terminal Information System (ATIS) radio broadcast that is updated whenever there is a significant change in conditions. Before and regularly during flight, pilots update the altimeter's pressure setting.

Attitude

The gyroscope in the **attitude indicator (artificial horizon)** is mounted in its gimbals so that it can be uses as a reference for both pitch and roll (see Figure 6-12). The instrument face has blue and black halves to indicate the sky and Earth, respectively. The artificial horizon separating them matches the location of the real horizon even when the pilot cannot see anything outside of the cockpit (see Figure 6-13).

Figure 6-12 The attitude indicator or artificial horizon uses a gyroscope to indicate the bank and pitch of the aircraft compared to a straight and level attitude.

© Image by X-Plane

Figure 6-13 Inside the artificial horizon is a gyroscope that is able to twist in two different directions. These motions are translated into a movement of the tan and blue "Earth" on the instrument face.

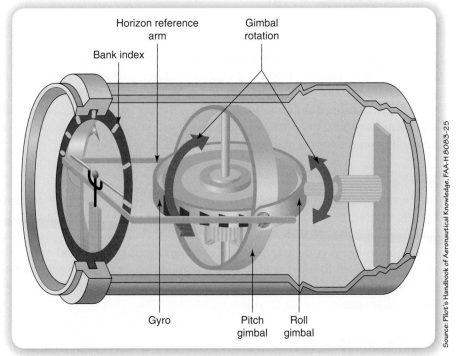

Horizon reference arm

Gimbal rotation

Bank index

Gyro

Pitch gimbal

Roll gimbal

Source: Pilot's Handbook of Aeronautical Knowledge, FAA-H 8083-25

Figure 6-14 The vertical speed indicator shows the rate at which altitude is changing based on how quickly the static pressure is changing around the aircraft.

Vertical Speed

The **vertical speed indicator (VSI)** contains a sealed chamber inside the instrument housing (see Figure 6-14). The housing has a calibrated leak so that its interior can come into equilibrium with the static pressure surrounding it (see Figure 6-15). If the static pressure decreases around the chamber, such as would happen during a climb, then the chamber expands to indicate the climb. The air inside the chamber

© Image by X-Plane

Figure 6-15 Inside the vertical speed indicator is an aneroid that is connected to the static line. A calibrated leak forces the instrument housings pressure to lag behind the instantaneous static pressure with the resulting pressure difference being proportional to the rate of change in altitude.

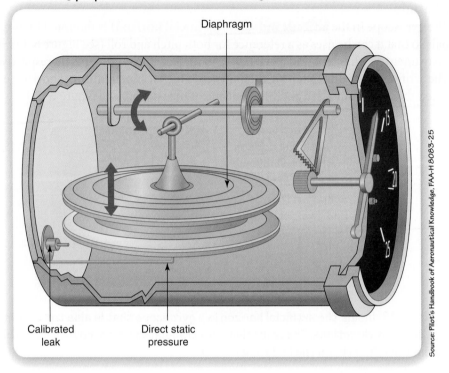

Diaphragm

Calibrated leak

Direct static pressure

Figure 6-16 The heading indicator uses a gyroscope to indicate the current heading of the aircraft during flight maneuvers. This gyroscope does suffer from precession errors requiring pilots to frequently check and adjust its indication to match magnetic headings.

begins to leak out in proportion to the rate at which the static pressure is decreasing. If the climb is ended, the chamber comes into equilibrium with the surrounding pressure, and the needle indicates the rate of climb or descent as zero. Some caution is required for both the VSI and the altimeter because flying along a constant pressure slope can guide you either upward or downward depending on the weather conditions.

Heading

Inside the **heading indicator**, a gyroscope is mounted so that the airplane's yawing motion is displayed by turning a card with magnetic heading indications (see Figures 6-16 and 6-17). The device is very accurate during maneuvering flight, unlike a compass; however, it suffers from precessional errors that must be corrected for during flight. On a regular basis, the pilot will adjust the heading indicator to match the compass heading while the aircraft is in straight and level, unaccelerated flight.

Turn Coordination

The gyroscope in the **turn coordinator** indicates the roll position and rate of turn of the aircraft while an inclinometer indicates whether or not the airplane is in coordinated flight (see Figure 6-18). Coordinated flight is the condition in which the airplane is yawing correctly to match the rate of turn. Thus, the instrument gives an indication of the roll attitude of the aircraft and whether or not sufficient rudder is being used to overcome adverse yaw (see Figure 6-19).

Figure 6-17 Inside the heading indicator is a gyroscope that is free to rotate around the vertical axis. This rotation is mechanically transferred to the face of the instrument.

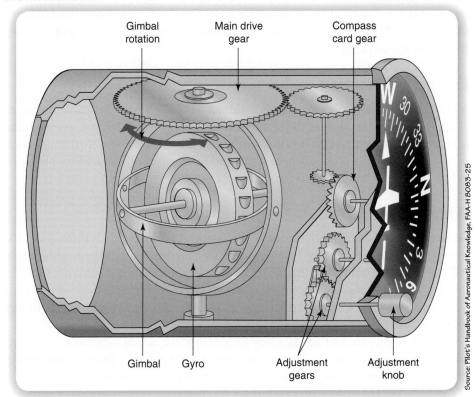

Gimbal rotation — Main drive gear — Compass card gear — Gimbal — Gyro — Adjustment gears — Adjustment knob

Source: Pilot's Handbook of Aeronautical Knowledge, FAA-H 8083-25

Figure 6-18 The turn coordinator uses an internal gyroscope to indicate the bank angle of the aircraft. The inclinometer uses a ball to indicate whether the aircraft is slipping or skidding through the turn. Pilots are taught to "step on the ball" to maintain coordinated flight.

© Image by X-Plane

Figure 6-19 Inside the turn coordinator is a gyroscope the twists around the longitudinal axis to indicate the roll attitude of the aircraft.

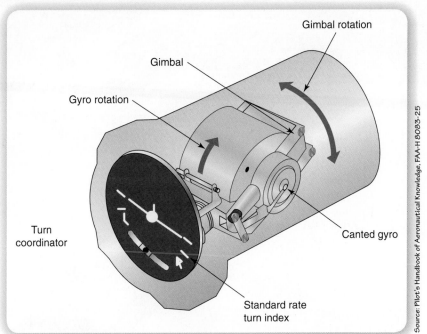

Gimbal rotation — Gimbal — Gyro rotation — Turn coordinator — Canted gyro — Standard rate turn index

Source: Pilot's Handbook of Aeronautical Knowledge, FAA-H 8083-25

Time

This is the plus one instrument. An accurate clock is critical to navigation as the turn coordinator is designed to help the pilot fly a standard rate turn. The standard rate turn is 3° per second or 2 minutes for a complete 360° turn. Pilots need to know the turn, climb, or descent rate, and the time over which they occur, to accurately maneuver the airplane.

Engine Information

Although not essential for navigation, almost all aircraft have instrumentation to indicate the RPM, as well as the cylinder head temperature, exhaust gas temperature, and other data related to the operation of the engines. This instrumentation is an important part of an efficient flight and can help diagnose and fix problems before they become a threat to the success of the flight.

Wind

Remember our 1-hour flight due east? We overlooked the fact that throughout the flight there was a 10 mph wind blowing directly from the south. Where in the world will we be at the end of the flight? There is almost always some horizontal movement to the air. These winds aloft vary in both strength and direction with altitude and can range from a minor annoyance to a life-threatening impact on flight (see Figure 6-20).

Therefore, the cockpit needs a way to monitor our actual location in comparison to our intended course and a way to correct for the winds. Doing so minimizes flight time and fuel consumption while maximizing range and safety.

GRASS STRIP ADVENTURE

Explore the mystery of a lost airliner named *Stardust* and a mysterious Morse code transmission of the word "STENDEC" at *www.pbs.org/wgbh/nova/vanished/*.

Figure 6-20 Air masses take on the characteristics of their source region. Air masses are moist or dry (maritime or continental) and warm or cold (Tropical or Polar).

From AHRENS. Meteorology Today: An Introduction to Weather, Climate, and the Environment, 9e. © 2008 Brooks Cole, a part of Cengage Learning, Inc. Reproduced with permission. www.cengage.com/permissions

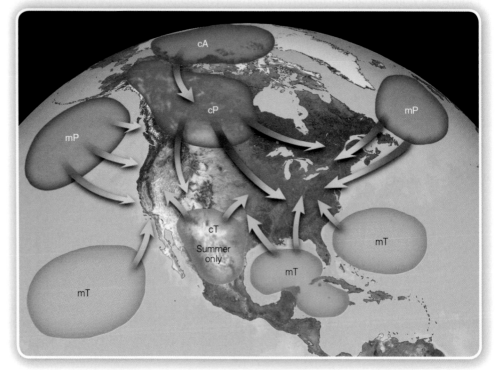

NAVIGATION AIDS

No matter what the instruments tell a pilot about the flight of the airplane, its navigational accuracy is always limited due to the fact that the airplane flies relative to the wind. From the cockpit, it is very challenging to get an accurate indication of how the air in which we are flying is moving. Starting in the 1920s, the aviation industry began to develop means to improve the ability to fly in a straight line from one point to another and to safely find airports in inclement weather.

Air Routes, Towers, Beacons, and Paint

An air route is nothing more than an official designation of the line along which airplanes should fly to get from one place to another. The first air routes were based on the transcontinental path from the East Coast to the West Coast. The expansion of the air route system mirrors the growth of the airmail and passenger services of the airlines.

Many of the early routes are based on already existing pathways used by pilots navigating by pilotage. Many air routes followed roadways, rail lines, or river valleys. To improve safety, these routes were augmented with aids to visual navigation. Along the route, the government installed sheds with painted roofs numbered with their location in the route (see Figure 6-21). Many provided paved or painted arrows on the ground pointing to the next station in the line, and some included rotating beacons to guide the pilot toward the station in low visibility. During World War II, throughout the United States, there was an initiative to have lettering and arrows painted on rooftops to guide our aircraft toward the nearest airport if an emergency landing was required.

Figure 6-21 A navigational shed and beacon from the early 1900s. Pilots would have to navigate from beacon to beacon to stay on the air route and safely arrive at their intended destination. A station was required about every 10 miles.

Airfield Markings

When an aircraft is approaching a paved runway, all of the markings and approach lighting has been standardized. Runway edges are illuminated with white paint and lights; the near end of the runway has green threshold lights while the farthest end has red lights (see Figure 6-22). Rotating beacons flash a color code of white, green, or yellow light to communicate where air, sea, or heliports are located. Standardization makes every airport somewhat familiar and easier to navigate when operating far from home.

Navigational Charts

Climbing into the front seat of a 1920s airmail airplane and flying was a very dangerous enterprise in the early days of aviation. One of the expert seat-of-the-pants pilots of the era was Elrey Jeppesen. Elrey developed a system of writing down notes about every new airport and its surroundings that he visited. He would describe the runways, the

Figure 6-22 Airport markings and lighting are standardized across the world so that pilots can fly into any active airport and recognize where their aircraft can and cannot safely operate.

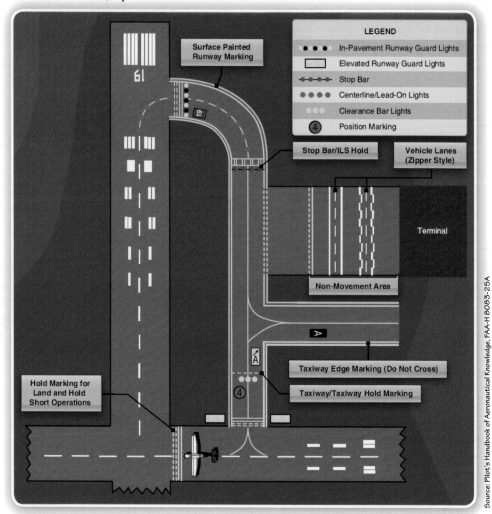

facilities, obstacles to navigation, and how to get the attention of the constable so they would come out to provide a ride into town.

Today Jeppesen-Sanderson is the world's largest provider of navigational and approach charts. Pilots around the world routinely use the detailed charts to plan and execute their approach to landing (Figure 6-23). On modern digital multifunction displays, a pilot can instantaneously pull up the relevant chart for every phase of flight. Access to charts specifically designed to solve the problems of aviation were a great leap forward from using roadmaps in the cockpit.

A/N beacons, NDB, VOR, ILS

The greatest enhancement to our ability to fly in extreme weather was the invention of electronic aids to navigation. Based on transmitters and receivers, radio-based aids to navigation allow pilots with a view from the cockpit that is completely obscured by the weather to "see" pathways through the sky that can guide them in a straight line toward their destination or landing site.

Figure 6-23 Sectional navigation charts are published by the Federal Aviation Administration. They include all the information required to safely navigate through an airspace except current weather and airport conditions, which the pilot obtains during preflight planning and during the flight.

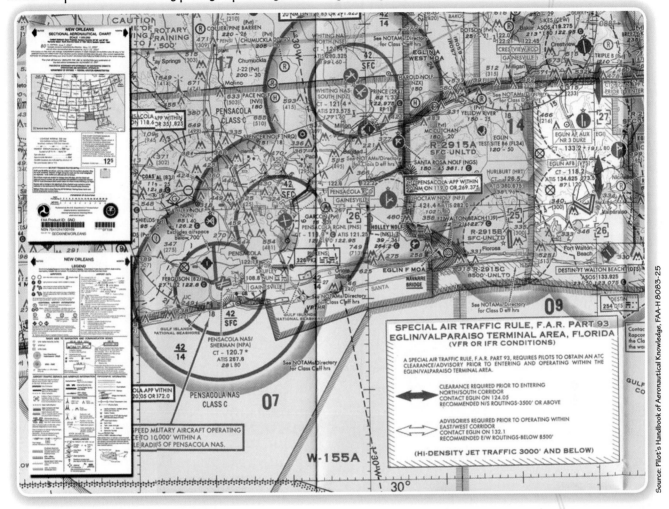

Source: Pilot's Handbook of Aeronautical Knowledge. FAA-H 8083-25

In the earliest days of radio navigation, a simple system of antennas broadcasted a simple Morse code A or N outward in a cone that is slightly wider than 90°. Working around the compass, the other letter was broadcast outward at 90° intervals. The A/N radio range system created four zones that overlapped at the edges with each of the A zones having an N zone on either side, and vice versa (see Figure 6-24).

Why a Morse code A and N? In Morse code, an A is coded as a dot-dash (• -) while the N is coded as a dash-dot (- •). If the pilot's aircraft was solidly in one of the zones, all that pilot heard through the radio was a single letter repeating. However, if the pilot was located in the overlap zone, the dots and dashes overlapped and produced a steady tone. Thus, early pilots would "fly the tone," or "beam," to the next station. If they drifted left or right, the tone became a Morse code letter, the letter indicating which way they had drifted off course.

Adding to this system, early aviators also used a radio direction finder to lock on to existing AM radio stations as well as stations built specifically by the government

Figure 6-24 *The LF/MF radio network creates a four-spoke radio navigation system to aid pilots of the late 1920's as they flew from station to station following the beam, or tone.*

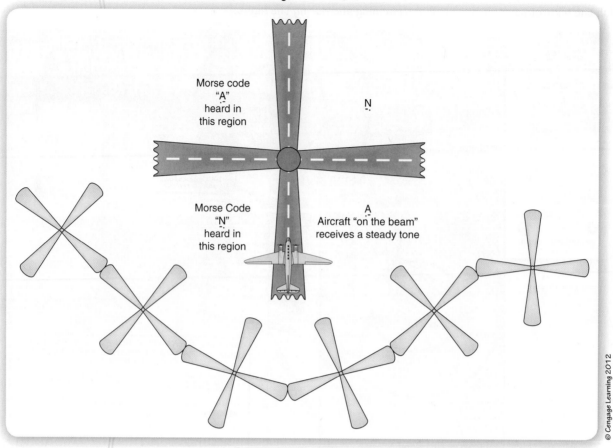

Morse code
"A"
heard in
this region

N
.-

Morse Code
"N"
heard in
this region

A
Aircraft "on the beam"
receives a steady tone

© Cengage Learning 2012

to fill in the gaps between stations along the needed routes. These **non-directional beacons (NDB)** offered only a heading directly toward the station, but that was enough to guide the plane to the station where it reoriented itself for continuing on its way.

A more expensive but vastly improved system of navigational aid is represented by the **Very high frequency, Omnidirectional Radio beacon (VOR)**. The VOR is a transmitter that sends out two signals simultaneously. One signal is pulsed outward in all directions (omnidirectional) every time that a rotating signal passes through the direction of magnetic north.

The two signals arrive at the same moment in time for anyone located directly north (magnetic) of the station. Everywhere else, the omnidirectional signal shows up first. How much later the rotating beacon signal arrives is directly related to how far around the compass circle the aircraft is currently located. In the aircraft, the VOR instrument monitors these two signals and converts the phase shift (difference in arrival time) of the two signals into a needle indication showing where any of 360° radial beams, each 1° wide, is currently located compared to the aircraft's current location. Using two VOR signals, an airplane can triangulate on a navigational chart to find its location to within 1° degree of accuracy. This system is easily accurate enough to guide an airplane to within visual sight of an airport.

Very high frequency, Omnidirectional Radio beacon (VOR):

a navigational aid that uses the phase shift between an omnidirectional and rotating beacon to create individual 1°-wide paths that radiate from the transmitter.

Point of Interest

Why Magnetic North?

When every other navigational aid fails due to a loss of signal or power outage, the remaining certainty is that Earth continues to create its magnetic field. To make certain that every system of navigation acts as a backup for the other means of navigation, all systems are based on this rudimentary way for defining a direction.

Because Earth's magnetic field changes over time, new navigational charts are published on a regular basis to show the updated magnetic course for any flight path (see Figure 6-25). In fact, FAA charts expire every 56 days!

Figure 6-25 The Earth's magnetic north pole (MN) is located hundreds of miles away from the geographic north pole (NP). The actual location changes from year to year requiring the frequent updating of all aerial navigation charts.

From AHRENS. Meteorology Today: An Introduction to Weather, Climate, and the Environment, 9e. © 2008 Brooks Cole, a part of Cengage Learning, Inc. Reproduced with permission. www.cengage.com/permissions

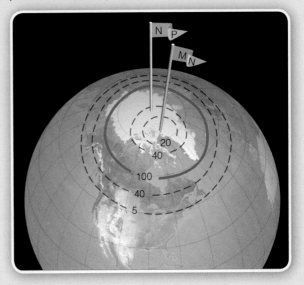

Distance Measuring Equipment (DME) works in conjunction with the VOR to provide the line of sight, or slant range, distance from the station to the aircraft. Using this equipment, a single station's signal allows pilots to find their location and navigate to an airport.

Many airports provide an enhanced radio navigational system for the approach to a specific runway called the **Instrument Landing System (ILS)**. An ILS system creates two elongated beams (see Figure 6-26). The first creates a vertical zone that is transmitted directly outward along the approach path for the runway. This beam helps the pilot maintain a flight path that aligns with the centerline of the runway. The second beam creates a horizontal radio zone that creates a sloped path upward from the touchdown location on the runway along the appropriate glide slope for the runway. This helps the pilot control the descent rate to fly a straight-line path to touchdown that is sufficiently high to avoid obstacles but as shallow as possible to minimize the approach speed of the aircraft. The cross hairs created by the crisscrossed beams thus guide an aircraft all the way to the point of touchdown; however, they don't create an awareness of how close the aircraft is to the runway threshold.

Figure 6-26 The Instrument Landing System (ILS) provides the pilot in the cockpit with information regarding the location of both the glideslope to the touchdown zone and the centerline of the runway. The pilot should fly toward the indicator needles to move closer to the correct flight path for approach and touchdown.

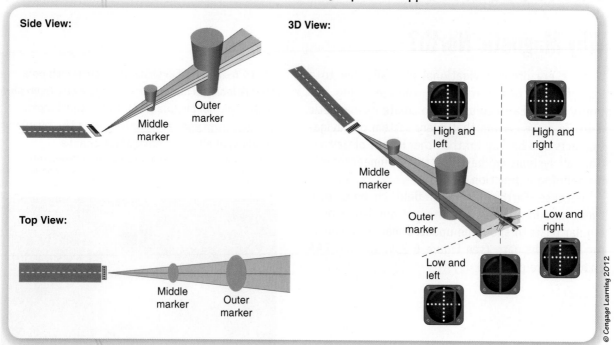

© Cengage Learning 2012

Minimum decision altitude (MDA):

the lowest altitude to which an aircraft can legally descend without making visual confirmation of the aircraft's position in relationship to a number of landing aids such as approach lighting, threshold lights, or runway lights, among others.

The last component of an ILS setup is a set of three transmitters that shoot a beam straight up into the air at different points along the approach path. These beacons are the outer, middle, and inner markers that identify when the aircraft is 3.9 NM (nautical miles) from the runway, at minimum decision altitude (MDA), and at imminent arrival for the approach, respectively.

Approach Lighting and Minimum Decision Altitude

All landings in the United States are required to be made while able to visually identify and confirm that the airplane is actually on the approach path and will touch down on the runway at the correct location. Instrument approaches have a published **minimum decision altitude (MDA)** at which the pilot in command has to have a visual fix on the landing, or the pilot is required to declare a missed approach and go around for another attempt at landing.

Pilots talk about an "approach to minimums" when they describe a situation in which the weather prevents the pilots from seeing any visual guides to landing until the aircraft is at the MDA. Pilots can continue on the approach if they can spot only the approach lights, the runway threshold, or runway side lights, or several other features.

RADAR CONTACT

Radio Detection and Ranging (RADAR) technology was developed during World War II. The system involves transmitting a radio signal and listening for the echo (Figure 6-27). Based on the travel time to and from the object causing the echo, and the direction in which the signal is being transmitted, a radar screen can paint

Figure 6-27 A microwave pulse is sent out from the radar transmitter. The pulse strikes an object and a fraction of its energy is reflected back to the radar unit, where it is detected and displayed.

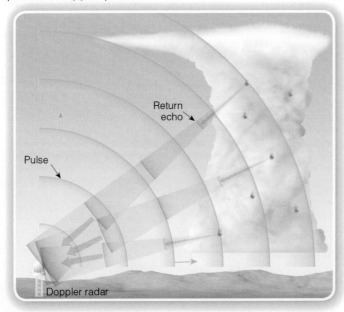

Figure 6-28 Senior Airman Elizabeth Hand of the Royal Air Force staffs the radar console at Lakenheath, England.

Courtesy of United States AirForce

an image of where objects are located. The system was limited in resolution and sensitivity but allowed the allied British forces to gain a significant advantage when combined with airborne radio equipment. Together, these technologies allowed for radio-based detection of incoming enemy forces, radar guidance toward enemy aircraft, and guidance back to safe airfields upon the return from a mission. Today, a heavily updated version of the early radar systems is employed for several technologies, not just aircraft. Modern radar systems, like the viewing station in Figure 6-28, employ higher resolutions scanning and can detect objects the size of large birds. Figure 6-29 shows a modern-day weather example of similar radar technology.

Figure 6-29 A satellite and radar image of Hurricane Humberto in 2007. Precipitation rates (lowest in blue, highest in red) were obtained by the satellite's Precipitation Radar and Microwave Imager. The rainfall estimates are overlain on the infrared image of the storm.

GRASS STRIP ADVENTURE

See 24 hours worth of air traffic on route into and out of Atlanta-Hartsfield International airport at *http://maps.unomaha.edu/ AnimatedFlightAtlas/atl.avi*.

Radar has matured into a sophisticated network that blankets the United States and most of the developed world; however, it is still limited in range, and much of the globe remains unprotected by the navigational information it can provide to pilots. As air traffic has multiplied drastically over the past 70 years, radar systems must be able to detect and track multiple recreational, commercial, and military craft (see Figure 6-30).

Radar and radio communication equipment provides the technological backbone for the modern air traffic control system. Pilots flying instrument flight plans can be guided from departure to arrival from a radar room located hundreds of miles away. In addition, radar controllers can warn pilots about severe weather

Fun Fact

Radar can be used to measure the distance to any object. Since 1974, Ground Proximity Warning Systems (GPWS) have become standard equipment on commercial jets. A GPWS uses a radar altimeter to trigger an audible "Pull Up, Pull Up" alarm when controlled flight into terrain is imminent. Since implementation, these systems have reduced the number of this type of accident from 10 per year to less than 1 per year.

Figure 6-30 The air traffic control system has experienced exponential growth since its beginnings.

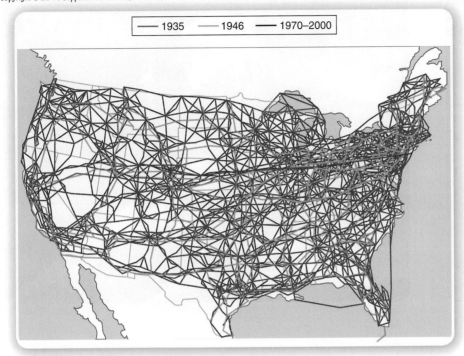

and provide guidance around the weather system. Controllers can also "call out traffic" to warn pilots about other aircraft in their vicinity. The system of radar and radio communication allows for the incredible number of flight operations that take place on a daily basis around the major cities of the world.

GLOBAL POSITIONING SYSTEM (GPS)

Starting early in the 20th century, scientists were proposing systems that used satellites as the radio station for navigation on a worldwide basis. After the 1957 launch of Sputnik by the Soviet Union, the reality of a satellite system became feasible. In 1978, the first of more than 24 GPS satellites was launched into orbit ushering in a world of precision navigation to any one with the proper equipment (see Figures 6-31a and 6-31b). However, the system remained accessible exclusively by U.S. military forces until after 1983, when Korean Air Lines flight 007 accidentally entered Soviet airspace and was shot down. The flight had many U.S. civilians and a congressman onboard. The outcry following this incident prompted President Ronald Reagan to make the system available for free to the global public to improve navigation and safety.

Challenge

It is easy to transfer information via radio if the transmitting and receiving equipment is large, and the energy is focused into a narrow beam. The GPS, however, is based on the concept that one constellation of orbiting satellites can provide meaningful data to an infinite number of receivers located anywhere on Earth. The challenge for the engineer was to find ways to improve the signal strength while miniaturizing the receiver and improving the accuracy of the final calculated location in real time.

Global positioning system (GPS):

a network of space satellites and receivers capable of accurately determining the location of the receiver anywhere and anytime on the surface of planet Earth.

Figure 6-31 (a) A GPS satellite in Launch.

Courtesy of U.S. AirForce

Figure 6-31 (b) A GPS satellite in orbit.

Courtesy of NASA

Constellation and Orbits

GPS is based on at least 24 satellites located in orbits that pass above the vast majority of Earth's population. The system includes a number of additional satellites to act as spares to ensure that the system is available 24/7 anywhere U.S. interests require navigational aid. The orbits are arranged so that 6 satellites occupy each of 4 orbital planes. Each satellite orbits at approximately 13,000 miles above Earth's surface completing its passage twice per day. This arrangement ensures that 6 satellites are always in direct **line-of-sight** from almost anywhere on Earth at any moment in time (see Figure 6-32).

> **Line-of-sight:**
>
> unobstructed—by walls, earth, or other objects— pathway between two objects.

Receiver and Data

Modern GPS receivers are very small and can simultaneously monitor 12 satellites while running on small batteries. The marvel of this technology breaks down into three distinct aspects of GPS and the signals used to transmit the information from the satellite to the client in an airplane. These characteristics are pseudo-random encoding, almanac information, and ephemeris data.

Pseudo-random code allows the receiver to be very small and still identify the satellite signal easily. Because all GPS data is digital, it is a series of 1s and 0s. When a GPS receiver is first powered on, it looks into its memory to estimate which GPS satellites should have line-of-sight to the location at which the unit was powered down but at the current time indicated by its cheap, relatively inaccurate, internal clock. If the antenna on the receiver is very small, the signal detected will contain a lot of noise.

Assume for a moment that there is not a GPS satellite in the sky and all that is received is noise. Noise is by definition random in nature. To look for a repeating 30-second block of information, we simply add each new block of 30 seconds of

Figure 6-32 *The GPS constellation of satellites is sufficient to ensure that at least 6 are visible from any location in the mid-latitudes at all times.*

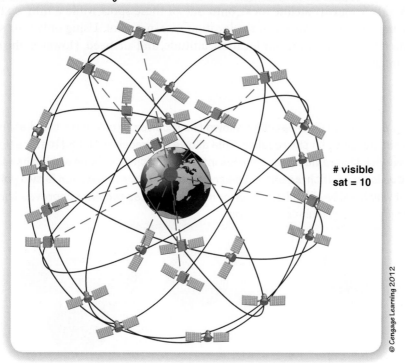

# visible
sat = 10

© Cengage Learning 2012

data to the existing sum. If the signal is truly random, the sum of all the blocks will remain very close to zero. However, if there is a very weak signal buried in the noise, then the sum that is the result of the noise will still be zero, but the part of the sum due to the occasionally received signal will slowly drift toward a 1 or a 0. This allows even very weak signals to be detected, and the portable, miniaturized GPS becomes a reality.

Almanac data allows the satellite to transmit the orbital characteristics for every satellite as well as corrections for signal degradation due to the ionosphere and time encoding down to the GPS receiver. Ephemeris data describes the precise location of any given satellite at every moment of time. This, along with the almanac and pseudo-random code, combines to allow the portable receiver to know exactly where and when each satellite's signal was transmitted. So how is this translated into a location?

Where Am I?

The basic function of a GPS receiver is to measure the time of arrival for each satellite signal and then to calculate a distance to the satellite based on the signal's travel time and the speed of light. This is an application of the basic physics equation:

$$Distance = Rate \times Time.$$

One satellite signal allows the receiver to place itself someplace on a very large spherical shell that is centered on the satellite and has a radius equal to the distance to the satellite calculated by the receiver. Two satellites allow the receiver to know that it is somewhere on the circle formed by overlapping the spherical surfaces from the two satellite signals. Add a third satellite to the mix, and the circle is reduced to two spots, the resulting intersection of three spherical distances (see Figure 6-33). A fourth satellite locks onto one of these spots while a fifth, sixth, and more improve the accuracy of the location measurement. Using only four signals, we can reliably determine our location, altitude, and airspeed. However, the system has limitations.

Figure 6-33 Each GPS satellite allows the receiver to calculate the distance from the satellite to the receiver. This allows the receiver to find its location on a sphere with this radius. Two spheres intersect with a circle on which the receiver must be located. If three satellites are received the intersection is two points, one of which will be ridiculous. Four or more satellites received improves the accuracy of the calculated location.

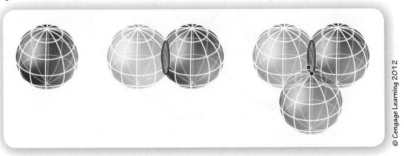

© Cengage Learning 2012

The microwave radio signals from the satellite travel at the speed of light, which is 186,000 miles per second. Each satellite contains its own atomic clock that is regularly adjusted to provide the time of transmission with accuracy good enough to calculate your current location to fractions of a foot. Imagine timing the arrival time of each satellite's signal with a watch that was 1 second later than the actual current time. For the first satellite, the calculated sphere would be 186,000 miles too large in radius. This is an error larger than the total distance to the satellite! The way for the low-cost clock in the receiver to calibrate to the atomic clock accuracy of the GPS satellite to be useful for navigation is in the math. The error in the internal clock's time is the same for every calculation. If the clock is late, the distance spheres from each satellite are too large, and if early, too small. All programmers had to do was have the receiver add or subtract very small amounts of time to each signal to make any four cross precisely at one location. Whatever amount of time the clock had to be shifted to achieve minimum error is the correction that is applied to the internal clock. For all subsequent calculations, the low cost internal clock is now as precisely timed as the very expensive atomic clock located on the satellite. GPS achieves thousands of dollars worth of performance for pennies, which is great engineering!

Improving GPS Accuracy

But we can do even better. The quality of the GPS calculated position is degraded by a number of different factors. To correct for these variables, the modern GPS has been augmented by the Wide-Area Augmentation System (WAAS) and differential GPS (see the following example).

Example

GPS Error Due to Technological Limits (Approximate)	
Calculation errors	1 meter
Clock errors on the satellite	2 meters
Ephemeris errors	3 meters
GPS Error Due to Atmospheric and Reflection Effects (Approximate)	
Ionosphere	5 meters
Troposphere	Less than 1 meter
Multipath Errors	Less than 1 meter

WAAS is based on the construction of a number of permanent, nonmoving GPS reference stations for which the actual location can be very accurately determined (see Figure 6-34). The stations calculate location correction data based on the real-time data being received from the constellation of GPS satellites. These corrections are broadcast within seconds up to a small number of geo-stationary WAAS satellites that broadcast the corrections to all WAAS-enabled receivers in the United States and most of Canada.

Figure 6-34 *The Wide Area Augmentation System (WAAS) reception area over North America.*

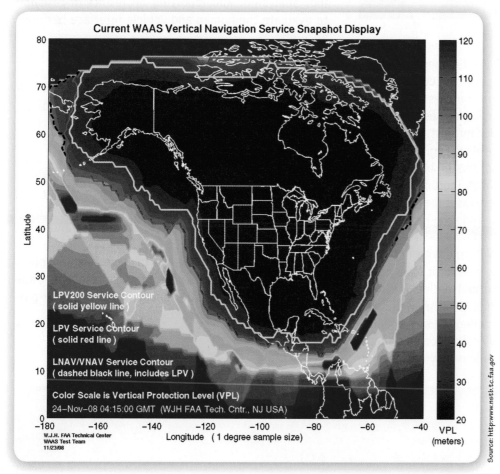

Careers in Aerospace Engineering

BUILDING THE HIGHWAY IN THE SKY: LYNDA KRAMER, NASA ENGINEER

Figure 6-35 Lynda Kramer is an aerospace engineer and principle investigator for synthetic vision systems at NASA's Langley Research Center.

Courtesy of NASA

Lynda Kramer's career at NASA Langley began in 1987 as part of the Engineering Cooperative Education Program. "I worked in six different branches as a co-op, which allowed me to figure out which research areas interested me," Kramer said. Her interest was heightened by her Bachelor of Aerospace Engineering degree at Auburn University and Master of Aerospace Engineering degree at George Washington University. While attending Auburn, Kramer was a member of the Auburn University Concert Choir and sang in the Soviet Union "while it was still under Communist control," she added. Kramer also sang in Poland with her choir. "I really enjoyed singing during a Catholic Mass at the former Pope's church in Poland and eating the most delicious ice cream I've ever had there as well," she said. "In my freshman year at college, I was a founding member of the Auburn University women's soccer club team," she said. She still enjoys playing soccer in her spare time. She also loves to spend time with her friends and family.

Some of Kramer's most memorable moments were "the birth of my daughter, Haley, scuba-diving to watch a shark feeding in the Bahamas, and tandem sky-diving." She strives to maintain a healthy balance between work and family. "NASA is very family-friendly, which makes this task easier to manage," she said.

Kramer has served for seven years as a principal investigator for synthetic vision systems and enhanced vision systems research, which are aimed at improving aviation safety and increasing aviation operations in restricted visibility. As an aerospace engineer, she specializes in crew-vehicle interfaces (CVI), with an emphasis on advanced displays that are expected to improve aviation safety and efficiency. "My technical duties include directing the design, development, testing, and overall integration of advanced displays into piloted workstations, flight simulators, and flight test aircraft, and also defining, conducting, analyzing, and reporting on results of research to evaluate the human and vehicle interface performance effects."

Kramer went through ground school for private pilot and instrument pilot training and rode jump seat aboard United Airlines revenue flights, "which helped me learn the operational world of the pilots that I'm creating the crew vehicle interfaces for," she said. It's satisfying for her to see NASA's CVI research successfully transferred to industry for multiple aviation applications. She hopes that these products will help "to improve aviation safety and potentially increase capacity." Her career decision that she made as a co-op has served her and aviation research well.

"I've been very fortunate to have been involved in cutting-edge crew vehicle interface research, while working with a fantastic, energetic, and highly productive team," Kramer said.

Navigation by WAAS provides accurate data 24/7 for category 1 instrument landings with no more than 5 minutes of downtime per year. The system improves navigational accuracy to within 1.5 meters in both location and altitude. Differential GPS is based on the same concept as WAAS but uses ground-based transmitters to communicate the corrections to the GPS receiver over a much more limited area.

Point of Interest

Automated Landings!

When the weather gets foul, aviators dream of the day when they can activate the automated landing system that will guide the airplane all the way to the ground, through the landing, and taxi them to the correct location on the airport ramp. The FAA breaks this dream down into steps progressively approaching the ultimate goal of automated landings for all aircraft.

Categories of Operation for Instrument Approaches to Landing			
Category Description	Minimum Decision Altitude (feet)	Runway Visual Range (feet)	Other
Category 1	200	2625	Visual enhancement technology allows proceeding to Cat 2.
Category 2	100	1200	
Category 3A	100 or none	700	
Category 3B	50 or none	150	
Category 3C Fully Automated Landing	None	None	Category 3C equipment can guide the aircraft through landing and taxi operations.

The Future

The GPS network continues to be improved with GPS III network capability adding ground stations, satellites, and four additional channels by 2013. These upgrades will enhance the system and allow a migration toward achieving the goal of category III instrument condition landings on a routine basis at any airport in the United States.

SYNTHETIC VISION AND SPATIAL AWARENESS

The next generation of navigational aids seeks to integrate data in all four dimensions of location and time into a visual display that is presented to the pilot in as natural a format as possible. Synthetic vision integrates three-dimensional terrain data, navigational aid databases, airport databases, and GPS information to calculate and present the environment surrounding the aircraft to the pilot in daylight quality regardless of the conditions (see Figure 6-36). Highway-in-the-sky systems can add simulated "hoops" in the virtual world of the synthetic vision system that allow the pilot to guide the aircraft from the start of its ground roll, through takeoff

Figure 6-36 Synthetic vision systems seek to integrate camera view, GPS, ILS, radar, traffic, and terrain data into a cockpit display that allows zero visibility operations of the aircraft as if it was in clear daylight conditions.

Courtesy of NASA

and landing, and back to its parking space on the airport ramp without ever visually identifying features in the real world through the airplane's windows.

Integrated Displays and the Glass Cockpit

Electronic displays and computer technology continues to evolve at a rapid rate. Since the turn of the century, it has become more affordable for new aircraft to ship with full glass panel multifunction displays (MFD) that can, with the touch of a button, access GPS, terrain, weather, approach charts, and many other data sources. Figure 6-37 is an early sample of the Space Shuttle's glass cockpit.

Daylight Approaches

The ultimate goal of a synthetic vision system is to transform nighttime and foul weather into a routine flight that can be reliably flown as if it was a daytime flight under ideal visual conditions (Figure 6-38). Combined with the improvements in accuracy in WAAS-based GPS and technologies under development, synthetic vision systems bring general and commercial aviation ever closer to achieving routine category III instrument landings.

Figure 6-37 Synthetic vision systems seek to integrate camera view, GPS, ILS, radar, traffic, and terrain data into a cockpit display that allows zero visibility operations of the aircraft as if it was in clear daylight conditions.

Courtesy of NASA

Figure 6-38 Synthetic vision displays allow nighttime operations as if it were daylight.

Courtesy of NASA

Careers in Aerospace Engineering

BREATH OF FRESH AIR

With the world's oil supplies dwindling, wind farms have been sprouting like mushrooms across the American landscape. Plentiful, renewable and remarkably clean, wind is an amazing natural resource, the gift that keeps on giving. But how do you get the electrical energy generated by a wind turbine to the designated end-user—say, a building in town? Put another way, why couldn't you design the building to harness its own wind power, generate its own electricity?

That's the project that Edward DeMauro, a PhD candidate at the Rensselaer Polytechnic Institute, is working on. "It's the aerodynamics of bluff bodies," DeMauro says, bluff bodies being things that are not meant to fly—blocks, cylinders, etc. "In my case, that means buildings. We're studying how air flows around buildings so that we can decide on the best places to put wind turbines."

What will those buildings look like, you might ask. "Right now, we have a ribbed cylinder called a 'WARP' design," DeMauro says. "Air speeds up between the ribs of the cylinder, and you can put small wind turbines between the ribs to speed up the air. This technology was invented back in the 1980s, and we're trying to improve on it for use in buildings."

Inspiration

Air flow wasn't the area DeMauro expected to go into while growing up in suburban New Jersey, between Princeton and Trenton. Airless flow was more like it—a rocket ship gliding through space. "I always imagined myself as a NASA engineer working on a probe that would explore new worlds," he says.

Early on, DeMauro thought about becoming an artist, but engineering kept tugging at his sleeve. "I'm one of those people who pays a lot of attention to detail," he says, "and my father, a statistician,

Edward DeMauro

was always pointing out the importance of mathematics. I remember falling in love with algebra in middle school but wondering what you could do with it. For me, engineering was the answer to the question of what to do with subjects like calculus, chemistry, and physics."

Education

That left the question of which engineering field to go into. "During my senior year of high school, everybody was having these epiphany moments—'Here's what I want to do with my life!'" DeMauro says. "All I knew was I wanted to do math and science."

He did them at SUNY-Buffalo where he earned a BS in mechanical and aerospace engineering and an MS in aerospace engineering. And how did DeMauro land on those areas? "The fact is, as a college freshman, you don't really know what an engineer is," DeMauro says. "I just looked down the list of all the programs and saw aerospace engineering, which seemed like the best chance of fulfilling my old dream of working for NASA."

NASA was the hook, but DeMauro makes a distinction between the hook and what becomes your niche. For instance, he found himself especially drawn to fluid dynamics. "In fluid dynamics, you start out with some really nasty-looking equations," he says, "equations that scare most people. I loved that challenge."

Advice

Not yet through with his own schooling, DeMauro is reluctant to hand out advice, but there's one thing he'd like to pass on. "The key is being diligent," he says. "You may not be the best at what you're doing, but just keep going. Studying engineering can be a steep upward battle, but there's a reward at the end."

SUMMARY

- Successful navigation in the early 1900s required a pilot to survive long enough to acquire a personal knowledge of the landmarks and obstacles along their air routes. Every time they were required to fly to a new destination, or when an emergency arose, it was a nerve-wracking and dangerous situation.

- By the 1930s, radio navigation aids were creating straight-line beams that connected the major airports by electronic highways in the sky. However, the network was thin in some areas and nonexistent in others. Aircraft needing to travel anywhere other than on the air route were right back to pilotage and dead reckoning.

- The invention of radar during World War II brought the opportunity to aid pilots in navigating anywhere within range of the system. However the signals are limited in range and are blocked by terrain. Radar provides the backbone for the modern air traffic control system and has dramatically improved safety by guiding aircraft and providing detailed information about current weather conditions.

- Precision approach systems have created finely tuned approaches to specific runways by providing pilots in the cockpit with a means to fly both the approach heading and descent glide slope to the runway's touchdown zone.

- Global positioning systems and synthetic vision systems provide the pilot in the cockpit with extremely detailed information about the current location of the aircraft and guidance to any location necessary for the safe operation of the aircraft.

- Aviation technology is quickly approaching the capability required for providing pilots with the capability for fully automated flight operations.

BRING IT HOME

1. Create a pilotage "flight plan" that describes how to travel from your home to school.
2. Using a website such as www.runwayfinder.com or Google Earth, create a flight plan appropriate to travel between two major cities in your region. Pay careful attention to the type of landmarks that will be visible from the air.
3. Hit the Internet to investigate radar transponders, ELT beacons, or Traffic Collision Avoidance Systems (TCAS), and how they aid in improving aviation safety.
4. Use Flight Simulator to fly regional flights using only pilotage, using only dead reckoning (reduce weather conditions to below VFR minimums), and then using map and/or GPS features as an aid to navigation.
5. Research the concept of "augmented reality." How is this different from synthetic vision? Where is this already used in aerospace? How can it be used to improve piloting a craft? How might it become a distraction?

EXTRA MILE

1. Use a GPS unit to travel a significant distance. During the trip, have passengers occasionally call out "flight diversions" to other unfamiliar locations. Navigate to one of the diversions using traditional maps and charts. (Pull over for safety while referencing the maps!) Navigate to another diversion using the GPS system and its built-in database of locations.

2. Research radio-direction-finding equipment. Contact a local wildlife researcher to demonstrate animal tracking equipment. Play a game of "Fox and Hound."

3. Learn Morse code, and practice with your friends.

4. Partner with a local pilot to use a flight simulator to learn how to fly VOR and ILS approaches into your local airport.

CHAPTER 7
Astronautics: It's Rocket Science

Before You Begin

Think about these questions as you consider the concepts in this chapter:

1 What is an orbit?

2 How does gravity affect travel beyond Earth's atmosphere?

3 How do we know where we are in space and how we are moving through space?

4 What do we need to know to predictably change the motion of an object in space?

5 What might Earth be like if we had a more elliptical orbit instead of our current nearly circular orbit?

6 How does the human element complicate space travel to other planets and stars?

7 What is the risk of Earth being struck by a large asteroid or comet?

Since the beginning of human awareness of the heavens, we have been fascinated by the stars. The earliest astronomers noted their motions as they rose from the eastern horizon and moved along curved paths toward the west. It wasn't long before they noticed that almost all of the stars behaved as if they were stuck on a crystal clear sphere that appeared to rotate around us. Among the stars, they also noted that a small number of objects seemed to wander across this sphere; these *Planetai* (Greek for wandering stars) had their own motions. Not until the 1600s did we begin to bind the heavens to the same laws of science that we could touch and feel and observe on the surface of Earth. From these roots was born the science of **astronautics**.

Astronautics:

the science and technology of travel beyond Earth's atmosphere, including orbital motion and flight between the planets, moons, and eventually the stars.

To understand our place in the universe, we must endeavor to be explorers (Figure 7-1). When we explore, we discover . . . sometimes that which we sought to uncover, but far more likely things that were unforeseen. The challenge then becomes to apply the newly gained knowledge to enhancing our lives through applications in science, engineering, and technological innovation.

Since the space race of the 1950s and 1960s, we have built an ever-increasing capacity to explore the solar

Figure 7-1 *Space Shuttle Endeavor returns to Earth after spending days orbiting it alone and docked with the International Space Station.*

Courtesy of NASA

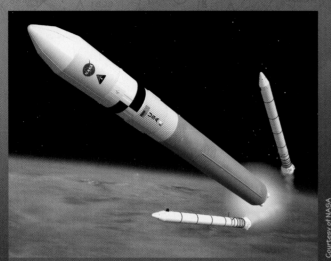

Figure 7-2 *The world's largest rocket booster, the Ares V, will be the primary heavy lift vehicle for human exploration missions to the Moon and Mars for the foreseeable future.*

Courtesy of NASA

system and the universe (Figure 7-2). But how do we move on a predictable path? What do we do when it's time to maneuver to a high orbit or initiate a reentry into Earth's atmosphere? And in very recent history, how do we change the orbit of a naturally occurring object? This chapter explores this knowledge and how it can be applied to any motion driven primarily under the influence of gravity.

WHAT IS AN ORBIT?

Orbit:
the curved path followed by an object as it moves around a planet or star or the act of moving along this curved path.

As early as the 17th century, we could describe how planets and orbits behaved in our solar system. We knew that an **orbit** was a path along which a planet moved to complete a closed path around another object. When they were first described, it was thought that all of the objects moved around Earth, which was thought to be stationary. We long ago realized that Earth is in motion around the Sun, and many objects orbit around objects other than Earth.

Fun Fact

Master Astronomer, Master Thief!

Kepler was an excellent astronomical observer and had access to the world's best observatory at Uraniborg, a tiny island off Denmark. The master of the observatory was Tycho Brahe, a renowned astronomer himself, who had designed the most accurate instruments for precisely measuring the angular positions of objects using the naked eye (telescopes wouldn't be invented until the early 1600s). On a daily basis, Kepler woke in the evening and observed all night, recording his data in his research journal, which every morning was locked away in Tycho's library. Kepler had no access to the data that he had collected.

It has been reported that, upon Tycho's death, Kepler requested access to his data but was denied repeatedly. Doing the only thing that he could think of, he broke into the library and stole the data! Although not the recommended course of action, this data allowed Kepler to recognize the geometric patterns of orbital motions that he later described as his laws of planetary motion.

Johannes Kepler (1571–1630) developed the laws that describe planetary orbits and motions. The patterns described in Kepler's three laws were later expanded on by Isaac Newton and his work with gravity and expanded further with the works of Albert Einstein and his theory of relativity, both of which provide explanations for why objects like the Earth, the Moon, the planets, and the stars move in the patterns that Kepler described.

Kepler's Three Law's of Planetary Motion

Let's look at Kepler's three laws: the law of ellipses, the law of equal areas, and the law of harmonies. As we describe these laws, we'll include some refinements made later by Newton and Einstein.

▶ **Law of ellipses.** The paths of the planets around the Sun are **ellipses**, with the Sun as one of the foci for that ellipse.

▶ **Law of equal areas.** A line drawn from the center of the Sun to the center of the planet will sweep out equal areas for equal amounts of time.

▶ **Law of harmonies.** The ratio of the square of the period (time) of any two planets is equal to the ratio of the cubes of their average distances from the Sun.

Kepler's First law: The Law of Ellipses

The first of Kepler's laws is probably one of the easiest to understand and is the foundation for the other two laws. First of all, an **ellipse** is a closed loop drawn so that the sum of the distances from any point on the curve to the two foci is constant (Figure 7-3). Ellipses belong to the family of math that deals with conic sections, or what is generated when a plane intersects a cone. Kepler's first law describes *how* planets behave geometrically, but it does not explain *why* they behave in this way. For the "why," we need some help from Newton and his observations on gravity. **Gravity** is the mutual attraction between any two masses. The closer the objects are or the greater the mass of either object, the stronger the gravitational force. Believe it or not, you and this book are gravitationally attracted toward each other. However, because you and the book both have a *very* low mass and reside on Earth—which is many magnitudes greater than the mass of you or the book—Earth's gravitational force is the only one apparent.

Ellipse:
the path followed by an object orbiting due to the force of gravity. A circle is a special case of an ellipse.

Gravity:
a field surrounding a mass that reaches to infinity but rapidly weakens with distance; the force that results from placing a second mass in the gravity field of another mass; the attractive force exerted equally but in opposite directions between any two masses.

Figure 7-3 An ellipse has two locations called foci, labeled A and B here. At any location on the ellipse, the total distance from both A and B is constant; that is, A + B = constant.

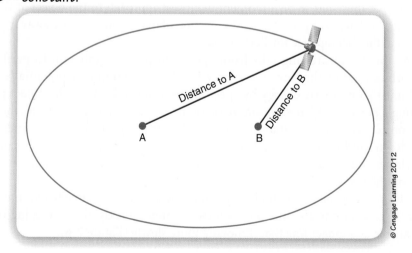

© Cengage Learning 2012

Newton's first and second laws do a great job of explaining Kepler's first law. Newton states in his first law that objects—in our case, planets—in motion tend to stay in motion, unless acted on by an outside force. Given that any two masses create a gravitational attraction between them, Newton's second law states that the orbiting planet will experience acceleration toward the center of the circle, called **centripetal acceleration**, because gravity—a force—is pulling the two massive objects together (Figure 7-4).

Figure 7-4 *The force of gravity and orbital velocity change during a typical orbit.*

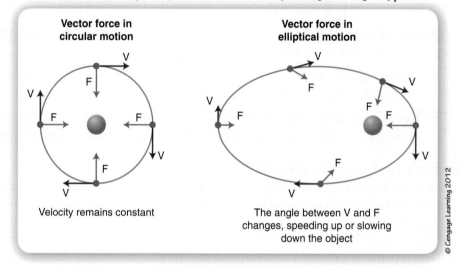

© Cengage Learning 2012

Kepler's Second Law: The Law of Equal Areas

Kepler's second law, the law of equal areas, states that for any unit of equal time (seconds, hours, days, years, etc.), the sweep of the area of a line drawn to the center of both objects during that time will be the same for any point on the orbit. This is easiest to see on a circle, which is a special type of ellipse. In a circular orbit, the planet is neither climbing away from nor falling toward the more massive object at the center of its orbit. When we bring Newton's gravity back into the picture, we can see that this means the object isn't losing or gaining gravitational potential energy, therefore, the orbiting object travels at a constant velocity. If the orbit is elliptical, then the object around which it is orbiting is off center at one of the two foci. This means that there has to be a moment of closest approach to the object at the foci (perigee for Earth at the center) and a location of maximum distance away from the object at the foci (apogee for Earth orbits).

As the object orbiting falls from apogee to perigee, aphelion to perihelion for orbits around the Sun, the object is losing gravitational potential energy and must gain kinetic energy by speeding up (Figure 7-5). As the object passes through perigee and maximum velocity, it begins to rise away from the central object, gaining gravitational potential energy by losing kinetic and slowing down until minimum velocity at apogee. For a two-body orbit, this pattern continues indefinitely with potential and kinetic energy repeatedly exchanging during the orbit.

The key to Kepler's second law is that as the length of the line between the two objects gets shorter, the line moves forward more quickly, thus sweeping out an equal area as a longer line that is moving more slowly (Figure 7-6).

Figure 7-5 This equation relates orbital diameter to required average orbital velocity.

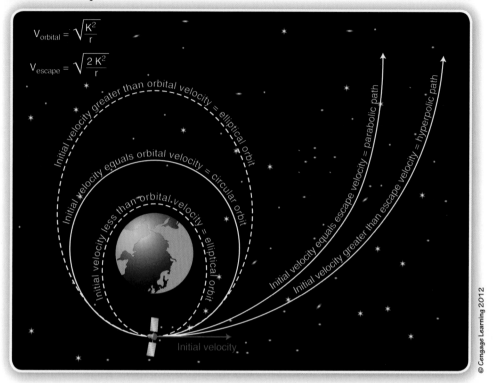

$$V_{orbital} = \sqrt{\frac{K^2}{r}}$$

$$V_{escape} = \sqrt{\frac{2K^2}{r}}$$

Initial velocity greater than orbital velocity = elliptical orbit

Initial velocity equals orbital velocity = circular orbit

Initial velocity less than orbital velocity = elliptical orbit

Initial velocity equals escape velocity = parabolic path

Initial velocity greater than escape velocity = hyperbolic path

Initial velocity

© Cengage Learning 2012

Figure 7-6 The law of equal areas. In any given unit of time, the line between the object being orbited and the orbiting object sweeps out an equal area.

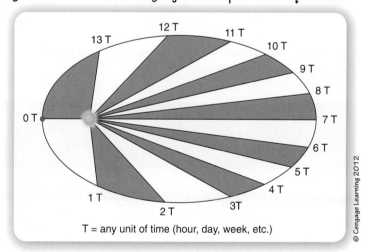

0 T, 1 T, 2 T, 3T, 4 T, 5 T, 6 T, 7 T, 8 T, 9 T, 10 T, 11 T, 12 T, 13 T

T = any unit of time (hour, day, week, etc.)

© Cengage Learning 2012

Kepler's Third Law: The Law of Harmonies

Kepler used mathematical terms to describe a pattern that connected the average distance between the two objects to how long it takes to complete one orbit. If you are farther away, it takes longer to complete an orbit. The mathematical relationship requires that the **period** (T) of the orbit squared divided by the average orbital distance (r) cubed should result in a constant value. The equation is $K = \frac{T^2}{R^3}$ (Equation 7.1) where

 K = the constant value for orbits around an object.

 T = period of the orbit, or time to complete one orbit.

 R = radius of the orbit.

For Earth's orbit around the Sun, we can determine the value of "K" from Kepler's math.

$$K = \frac{(1\ yr)^2}{(1\ AU)^3} = 1\frac{yr^2}{(AU)^3}$$

Using relative values for any other object, such as Mars, allows us to quantify this relationship so that if we compare Earth's orbital period of 1 year and its distance to the Sun of 1 astronomical unit (AU) to that of Mars at approximately 1.52 AU, we can respond with a numerical value. In this case, the distance multiplier is "1.52 x" the distance. We cube this result ($1.52^3 = 3.51$), and then take the square root $\sqrt{3.51}$ to find that Mars should take 1.87 Earth years to orbit the Sun or 684 days. The actual orbital period for Mars is 687 days. This should be expected because Kepler derived his mathematical pattern from observations of the planet's motion over time. But it leads us to ask whether this ratio of $K = \frac{T^2}{R^3}$ (Equation 7.1) actually is a constant for all objects in the solar system.

Fun Fact

Astronomers invented the astronomical unit (AU) and defined it as the average distance between Earth and the Sun so that it could be used as a simplified unit of distance for measuring things within the solar system.

Following are the results when the ratio is calculated for the planets within our solar system using the law of harmonies:

Planet	Period (yr)	Ave. Dist. (AU)	$K = \frac{T^2}{R^3}$ $\frac{yr^2}{(AU)^3}$
Mercury	0.241	0.39	0.98
Venus	.615	0.72	1.01
Earth	1.00	1.00	1.00
Mars	1.88	1.52	1.01
Jupiter	11.8	5.20	0.99
Saturn	29.5	9.54	1.00
Uranus	84.0	19.18	1.00
Neptune	165	30.06	1.00
Pluto	248	39.44	1.00

Nearly every planet has a third law ratio for period to distance close to 1. The same ratio and result applies to the Moon and artificial satellites.

Figure 7-7 The law of harmonies.

The Law of Harmonies suggests that the ratio of the period of orbit squared (T^2) to the mean radius of orbit cubed (R^3) is the same value k for all the objects orbiting the central body. For Kepler, all the planets orbiting the sun have one "K" value.

$$\frac{T^2}{R^3} = K \qquad \begin{array}{l} T = time \\ R = Radius\ of\ orbit \end{array}$$

If Earth revolves around the sun in one year and the distance from the Earth to the sun is 1 Astronomical Unit (AU), we get a value of 1 for K.

$$\frac{(1\ yr)^2}{(1\ AU)^3} = 1\frac{yr^2}{AU^3} = K$$

What about the other planets?

GRAVITY'S INFLUENCE ON ORBITAL MOTION

Once again, we turn to Newton and gravitational forces for a better description of what is going on. Recall that Newton developed the first law of motion (inertia) to define that the natural motion of an object is to keep doing what it is doing. A planet with a velocity should move in a straight line, not in an elliptical path, unless there is an outside force pulling on it. Newton hypothesized that the Moon is held in orbit by the force of gravity. But he needed more to prove that gravity was strong enough to create the centripetal force necessary to cause the orbit of the Moon that he observed. Newton, using Kepler's third law as a comparison, was able to combine his law of universal gravitation with circular motion to prove that gravity can match the centripetal force. From Figure 7-7, we used the unit ratio of years to astronomical units to get a value of 1. Using dimensional analysis, we can convert that ratio to $K = 2.97 \times 10^{-19} \frac{s^2}{m^3}$, which represents Kepler's constant, for the ratio of $\frac{T^2}{R^3}$.

The net centripetal force acting upon this orbiting planet is given by the relationship

$$F_{net} = \frac{(M_{planet} \cdot v^2)}{R} \qquad \text{(Equation 7.2)}$$

Where:

F_{net} is the net centripetal force.

M_{planet} = mass of the planet

v = velocity.

This net centripetal force is the result of the gravitational force, which attracts the planet toward the Sun, and can be represented as

$$F_{gravity} = \frac{(G \cdot M_{planet} \cdot M_{Sun})}{R^2} \qquad \text{(Equation 7.3)}$$

Because $F_{gravity} = F_{net}$, the preceding expressions for centripetal force and gravitational force are equal. Thus,

$$\frac{(M_{planet} \cdot v^2)}{R} = \frac{(G \cdot M_{planet} \cdot M_{Sun})}{R^2} \qquad \text{(Equation 7.4)}$$

Because the velocity of an object in nearly circular orbit can be approximated as

$$v = \frac{(2\pi R)}{T} \qquad \text{(Equation 7.5)}$$

$$v^2 = \frac{(4\pi^2 R^2)}{T^2} \qquad \text{(Equation 7.6)}$$

substitution of the expression for v^2 into the Equation 7.4 yields

$$\frac{(M_{planet} \cdot 4\pi^2 R^2)}{R \cdot T^2} = \frac{(G \cdot M_{planet} \cdot M_{Sun})}{R^2}$$

(Equation 7.7)

By cross-multiplication and simplification, Equation 7.7 can be transformed into

$$\frac{T^2}{R^3} = \frac{(M_{planet} \cdot 4\pi^2)}{(G \cdot M_{planet} \cdot M_{Sun})}$$

(Equation 7.8)

The mass of the planet can then be canceled from the numerator and the denominator of the equation's right side, yielding

$$\frac{T^2}{R^3} = \frac{(4\pi^2)}{(G \cdot M_{Sun})}$$

(Equation 7.9)

The right side of the preceding equation will be the same value for every planet regardless of the planet's mass. Subsequently, it is reasonable that the $\frac{T^2}{R^3}$ ratio would be the same value for all planets if the force that holds the planets in their orbits is the force of gravity.

Working with formulas

Deriving Newton's equations . . .
Because the position and velocity vectors move in tandem, they go around their circles in the same time T. That time equals the distance traveled divided by the velocity (Figure 7-8)

$$T = \frac{2\pi R}{v} \quad \text{(Equation 7.10)}$$

and, by analogy,

$$T = \frac{2\pi v}{a} \quad \text{(Equation 7.11)}$$

Setting these two equations equal and solving for a, we get

$$a = \frac{v^2}{R} \quad \text{(Equation 7.12)}$$

Figure 7-8 Uniform circular motion. Notice that if the object orbits at just the right speed, it moves in a circle. This means that the force of gravity and velocity vectors stay perpendicular to each other and the object neither speeds up nor slows down.

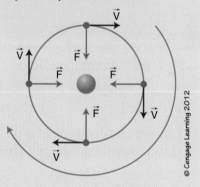

© Cengage Learning 2012

To help understand the interaction between forces and motions in an orbit, let's use an example of a tennis ball on a string being spun around your hand (Figure 7-9). Imagine your hand is the Sun, the tennis ball is Earth, and the connecting string represents the gravitational attraction between the two. If you spin the tennis ball slowly with your hand at the center of the rotation, you should notice something about the orbit of the tennis ball. The string is not straight out from your hand. If you increase the angular speed, the tennis ball speeds up and rises even with your hand. The speed of the tennis ball wants to rip outward away from your hand but the string won't let it, so it has no choice but to travel in a circle. There are two actions at work here. The velocity represents an inertial

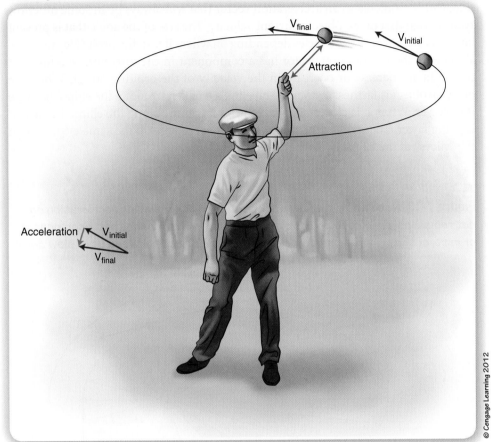

motion of the ball and its tendency to continue in a straight line, while the gravitational attraction (the string) between the two objects causes acceleration into a curved path. Both of these are independent from each other, meaning they are not related. The force is uniform for a circular orbit (remember gravitational attraction changes with distance from an object) because the radius is the same throughout the orbit; the velocity will be constant because the gravitational force is always perpendicular to the motion of the planet, toward the center of the orbit. Thus, gravity neither pulls forward nor backward on the planet, leaving its speed unchanged, but it does act to curve the speed to a new direction as the planet moves around its orbit.

Keep in mind that both the velocity and the acceleration of the ball are always changing because the string is applying a force toward the center of the orbit. This is a different direction at each location in the orbit. Remember, acceleration doesn't always mean an increase or decrease in velocity; it can also cause a change in only direction.

With constant velocity, it's easy to observe that the tennis ball creates an even arc and that the string sweeps out an equal area for any given second; whether it's at the top of the rotation or the bottom. This is also known as **uniform circular motion**.

But a circle is a special case, and no other elliptical orbits maintain a constant velocity. So, what gives? For help with the law of equal areas, let's approach it from Newton's point of view as forces and vectors. Remember that both objects have gravitational attraction, which is a force. You know that force is a vector, with both magnitude and direction. The other vector is the velocity of the orbiting object. In physics, we often add or subtract forces to get a resultant force. The force of gravitational attraction creates an acceleration vector that will always be in some other direction than that of the object's current velocity. The size of the angle that is present between these two vectors varies depending on where it is in the orbit (Figure 7-10). Whenever the acceleration vector has a component in the direction of velocity, it results in an increase or decrease in the velocity of the orbiting object. When the orbiting object is farthest away, it is traveling a smaller arc along the ellipse because it is slower. When it is closest, it is traveling a larger arc along the ellipse because it is going faster. The areas of the two opposite sweeps are still the same even though the "slices" of the ellipse look very different.

Figure 7-10 Notice how the force of gravity and the resulting acceleration do not remain perpendicular to the object's velocity. This means that a component, or part, of the gravitational force can cause the object to speed up or slow down.

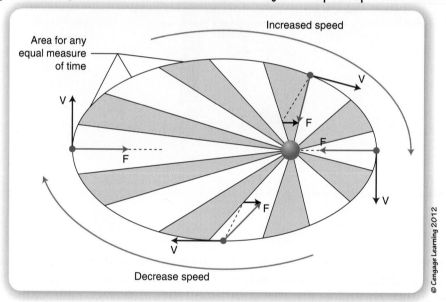

© Cengage Learning 2012

We can summarize this motion by applying some new terminology. The prefix "apo" comes from the Greek and means far or away from, while the prefix "peri" means close or near to. Applying these to the name for our Sun (*Helios*) and Earth (*Geos*), we can identify two unique locations in any noncircular orbit. For an orbit around Earth, these are **apogee** and **perigee**; for the Sun, they are **aphelion** and **perihelion** (Figure 7-11). As an object moves through closest approach, it is moving at its greatest speed. This speed is too fast for a circular orbit at its present distance, so it hurls itself outward from the object it is orbiting. However, this outward motion means that the force of gravity acts as a braking force and slows the object until it no longer can rise against the force of gravity. At this point, it is at its farthest and slowest point in the orbit, at which it doesn't have enough velocity to remain in this enlarged orbit and thus gravity begins to pull it inward. From farthest to nearest approach, gravity accelerates the object to higher velocity until it once again reaches its maximum at closest approach and the whole cycle repeats—almost identically for every orbit.

Figure 7-11 At the points of farthest (aphelion) and closest (perihelion) passage, the force of gravity is perpendicular to the velocity. For a brief moment in time, the velocity stops changing speed and only changes direction.

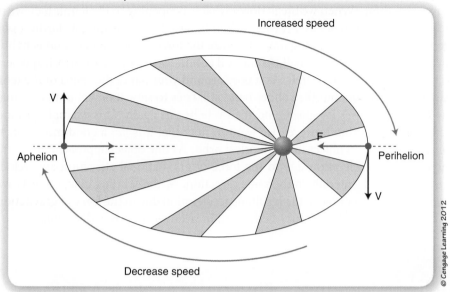

© Cengage Learning 2012

Fun Fact

Simplifying Space and Time

Realize that for the purposes of this book, we are simplifying orbital forces in terms of a two-dimensional plane and in the context of only two objects involved, the object orbiting and the object being orbited. In reality, satellites—manmade or natural—orbit around planets (or other satellites such as moons), and those planets in turn revolve around a star such as the Sun, which is moving and being influenced by a galaxy that is also moving through the universe in its own path. Multiple forces are acting on any given object at one time in space, even if you think you are far away from any objects (Figure 7-12). We are only concerning ourselves with the simple relationship between two objects at a time. If you would like to get more in depth with the more complex interactions of space and time, a career in astrophysics might be a good idea.

Figure 7-12 The beehive of human manufactured objects large enough to be tracked by radar from the Earth's surface. Notice how many of them are located in the geosynchronous orbital belt around Earth's equator. NOTE: Objects are NOT to scale and have been greatly enlarged to make them visible.

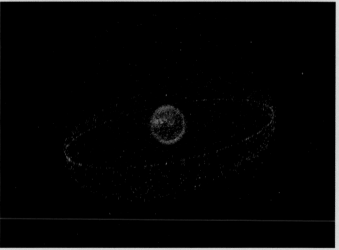

Courtesy of NASA

To use our tennis ball example for an elliptical orbit, imagine replacing the string with a large rubber band. Instead of spinning the tennis ball evenly, we try and spin it toward one direction more than any other. If we time it right, the ball travels in an elliptical fashion. When we break down what is happening to the tennis ball, we notice a change in velocity as it travels. When the ball is approaching its farthest point from your hand, it is slowing down because the force of the rubber band is fighting the forward velocity of the ball. As the ball continues on the return trip, it speeds up as the rubber band pulls on it. As it goes around your hand, the speed of the tennis ball is at maximum, but the rubber band still tries to slow it down, until eventually the tennis ball loses its speed, and the cycle begins again. This example produces a nice elliptical orbit; however, we want to note that it is *not* a good model for the forces acting on an object in an elliptical orbit. As one object approaches the other, the resultant acceleration due to the force of gravitational attraction and velocity are additive and generally in the same direction. As the orbiting object travels away from the stationary object, the resultant acceleration due to the force of gravitational attraction and velocity are subtractive and in relatively opposite directions.

Your Turn

Planetary Pencil Experiment
You can make an ellipse easily with a loop of string, two pushpins, and a pencil.

Materials
2 pushpins, pencil, loop of string, bulletin board or similar material

Steps
1. Push the pushpins into the bulletin board a few inches apart from each other.
2. Make a loop of string that is a little larger than the distance between the two pins. Put the loop of string around the two pins.
3. Put the pencil inside the loop of string and gently pull outward to tighten the loop. Move the pencil in a circular motion while keeping the string under the same tension.

For fun, try moving the foci closer together or farther apart. What do you get when both foci are in the same place (only one pin)? How could you calculate or predict the size of your ellipse?

Fun Fact

You Are Here—The Celestial Sphere

The Celestial Sphere
To help describe the location of objects in the sky, we use the **celestial sphere** (Figure 7-13). This sphere has an infinite radius but shares the same poles and equator as Earth. Imagine the earth being inside a large clear beach ball. The point on the sphere directly overhead for an observer is the **zenith**. If you were to draw an arc on the sphere that connected the poles with the zenith, this would be the observer's **meridian**.

Fun Fact *(Continued)*

Figure 7-13 **The celestial sphere with observer's position noted.**

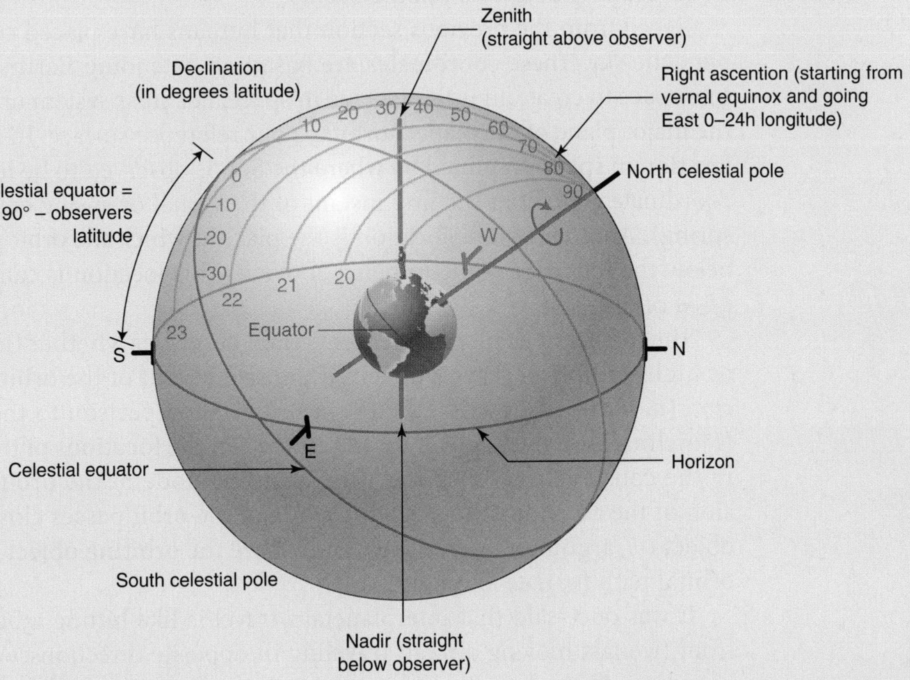

Declination and Right Ascension

Recall that on Earth to describe our geographic location, we often use latitude and longitude. Latitude refers to the rings that circle Earth starting from the equator and getting smaller and smaller as they reach the poles. Think latitude equals ladder. Longitude rings connect the North Pole to the South Pole and are all equal in diameter. Think longitude equals longer. In much the same way, we have latitude and longitude to give us coordinates on Earth: Declination and Right Ascension give us coordinates for celestial objects. Declination (DEC) is the "latitude" of the celestial sphere, and it is expressed in degrees just as it is for Earth. Instead of using north and south however, declination degrees are expressed as positive (+) for north or negative (−) for south. Because the celestial sphere shares the same poles and equator as Earth, the celestial equator is 0° DEC, and the poles are +90° DEC and −90° DEC. For celestial "longitude," we use the term right ascension (RA). Right ascension can be expressed in degrees, but it is more common to specify it in hours, minutes, and seconds of time. The reason? As we go through a day or 24 hours, the sky has appeared to rotate 360° (even though we are the ones who have rotated). So with some simple math:

$$360°/24hr = 15°/hr$$

Just as you would use latitude and longitude to describe a specific location on Earth, right ascension and declination can describe a specific position in the celestial sphere. Polaris, also known as the North Star because the Earth's northern axis appears to point at it, has coordinates RA 2:43.6 DEC +89.31

INTERPLANETARY TRAVEL

When we understand planetary motion and orbit mechanics, we can predict how objects will behave in space. If we can predict their behavior, then we can make the calculations necessary to guide the vehicle to other planets or objects.

To fully describe an object's orbital motion, we need a small set of coordinates, or numbers, that describe the geometry and current state of the orbit. These numbers are called **orbital elements**. A complete set of Keplerian elements contains six numbers, two that describe the size and shape of the orbital ellipse, three that describe its alignment and orientation in space, and one that identifies the location of the orbiting object within the orbit.

Recall from the previous section that humans have placed coordinate systems onto the sky. These coordinates are based on extending Earth's equator outward into space to create an infinite plane in space. For solar system orbits, we might use the major plane of the planet's orbits as our reference surface. In any case, a unique location in space is defined by where the Sun is observed to be located against our coordinate system on the first instant of the vernal or spring equinox (first day of spring). Another unique location is the place at which the orbit passes from being below the reference plane to being above it. This location is called the **ascending node** of the orbit.

The full set of Keplerian elements then describes whether the orbit is circular or highly elliptical (e, eccentricity), the size of half of the orbit's longest dimension (a, semi-major axis), the tilt of orbit in comparison to the reference plane (I, inclination), the angle of rotation between the locations of the vernal equinox in the constellation Aries and the ascending node of the orbit (Ω, right ascension of the ascending node), which side of the orbit passes closest to the central object (ω, argument of perigee), and where the orbiting object is located on this orbital path (v, true anomaly).

It was once said that interplanetary travel is like hitting a bullet with a bullet from two fast moving aircraft traveling in opposite directions. After the orbit has been completely described, we can use thrust forces to modify the velocity of the object to push it into a new orbit. In this way, we can make the orbit more or less circular, larger or smaller, and/or reoriented in space (Figure 7-14). Consider what would have to happen to make an orbit change so that it intercepts some place in space that you are interested in exploring. Of course, this is only half the battle of interplanetary travel. The other half involves building the technology that will get there most efficiently and survive.

Figure 7-14 Transfer orbits are used to move an object from one orbit to another by firing the engine to move either the point of farthest or nearest approach to match that of the new orbital radius. A second burn is required to make the orbit circular upon arrival.

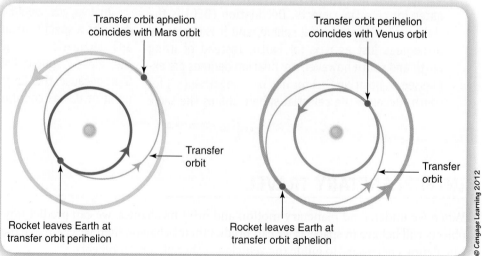

RENDEZVOUS IN SPACE: GETTING THERE WITH HOHMANN TRANSFERS

There are several options when trying to aim for objects in space. Because weight is such an issue when traveling into space (due to the amount of energy needed to put something in orbit), engineers try to minimize the amount of propellant needed for maneuvering in space. This means interplanetary rendezvous has to be creative and timely.

The most fuel-efficient method is to make a minimal burn required to change the orbit by timing it so that the new resultant orbit intersects and rendezvous with the new object. This method is known as a **Hohmann transfer** orbit.

In essence, it's much the same as a quarterback throwing the football to a wide receiver. The quarterback has to predict where the runner is going to be and release the football in such a way that the path of the football and the path of the runner cross the same point at the same time.

Hohmann transfers use an elliptical orbit that is tangent to initial and final orbits. Simply put, the method figures out a starting point and an end point and calculates an ellipse that shares those two points on its arc. This method uses very little fuel because most of the interplanetary travel is governed by orbital mechanics. Ideally, thrust is only applied in the same axis as current velocity; therefore, fuel is only used to accelerate a satellite into the transfer orbit and then slow it down as it approaches its destination. So for a successful Hohmann transfer to occur, the orbit must be in the same plane. This allows for the best approach to the object with minimum acceleration adjustment. The major axis of both orbits must align, and the velocity changes must occur tangent to the initial and final orbits. Hohmann transfers can move a satellite from Low Earth Orbit (LEO) to Geosynchronous Orbit (GEO) in about five hours, or into an orbit that intercepts the Moon in about five days. Travel to the planets requires much longer timeframes even for visits to our closest neighbors in space. Mars requires a transfer orbit of about 260 days! To visit objects much farther away, we often make use of gravitational slingshots to reduce the travel time, but we will come back to that in a moment (Figure 7-15).

> **Hohmann transfer:**
>
> a series of two rocket engine burns that initiate departure from a stable circular orbit around one object and entry into a stable circular orbit around a second object. An example is the transfer by the *Apollo* spacecraft as it traveled from Earth to the Moon.

Figure 7-15 A Hohmann transfer maneuver. From a circular orbit an initial acceleration hurls the object outward into an elliptical orbit. As it moves outward it slows until aphelion is reached. At that point, it would normally begin to return to the lower orbit, picking up speed along the way. However, a second burn at aphelion makes the final orbit circular at the enlarged radius of the new orbit.

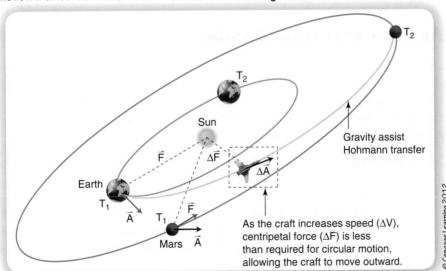

So how much energy must we add or subtract to change to a specific orbit? As we saw with Newton and Kepler, there is a relationship with the velocity of an orbiting object and the distance it maintains depending on the gravitational attraction of the two objects. So if we can find the velocity required for the new orbit, we just have to calculate the change in velocity required and the amount of time over which the engines will fire. From this, we can find the acceleration that has to be produced by the engines. Combine this with the mass of the spacecraft, and we can calculate the force that has to be produced by the engines. It might be rocket science, but the simplest rules of physics allow us to predict and control changes in the orbital dynamics of an object.

Gravitational Slingshots

We have already described how the force of gravity can be used to hold objects in a path that we call an orbit. However, gravity can also be used as the engine that supplies the force required to greatly reduce the time required to reach objects that are very distant in space. Gravitational slingshot maneuvers are generally used to drastically increase, or decrease, the velocity of the spacecraft and to change its direction of travel.

Speeding up is accomplished by passing behind the larger planet. The usual action of falling toward the planet will produce an increase in speed during approach and a slowing during departure from the planet; however, due to the orbital motion of the larger planet itself, we transfer a significant amount of energy from the larger planet's orbital motion to the smaller spacecraft. For the low-mass spacecraft, this appears as a large increase in velocity while the energy loss for the large planet is not measurable in the short term. By the same process, we can pass in front of the larger planet and drastically slow down the spacecraft.

So how do we use a gravitational slingshot to change the direction of the orbit? Imagine the head-on view that our spacecraft pilot might see out the front viewers of the spacecraft during approach to a massive planet. If the spacecraft's orbit intersects with the center of the planet, we would witness a collision as the planet loomed larger and larger in the viewer. However, if we just miss hitting the planet, we would be pulled "downward" by its gravity toward its center. Miss to the right, and our orbit is pulled hard left; miss to the left, and we are pulled hard right. But if we miss just over the top of the approaching planet, we are pulled strongly downward out of our old orbital plane. By controlling the approach angle to the planet, we can gain or lose overall velocity and steer the spacecraft toward a new place in space.

EARTH'S ORBIT AND SEASONS

Earth's orbit is very nearly circular, but it does have a place of closest approach (perihelion) and farthest approach (aphelion) from the Sun. This has led many people to assume that we experience the seasons of warmth and cold due to distance from the Sun, but that is not true. First let's clarify why distance from the Sun is not the primary cause of Earth's seasons:

▶ **Strike one:** There is one Sun and one Earth. This means that there is one distance between the two objects, so there would have to be one shared cycle of the seasons in which everyone on Earth would have the warmest and coldest days on the same days of the year. This does not occur. We know that northern latitude locations experience opposite seasons from those experienced in the Southern Hemisphere. Winter in the United States is summer in Australia (Figure 7-16).

Figure 7-16 A top-down view of Earth's orbit around the Sun. Note how a difference of millions of miles between perihelion and aphelion distances makes very little difference in how elliptical the orbit appears to be. Also note that Earth is closest to the Sun on January 4th each year.

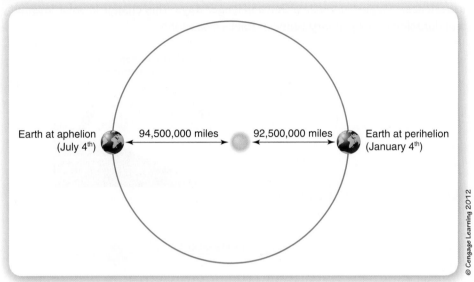

Earth at aphelion (July 4th)   ◄— 94,500,000 miles —►   ◄— 92,500,000 miles —►   Earth at perihelion (January 4th)

© Cengage Learning 2012

▶ **Strike two:** Northern Hemisphere summer occurs when we are farthest from the Sun. Farthest from the Sun day (the aphelion point in our orbit), occurs on July 4th. If distance causes the seasons, this ought to be winter. Note that Southern Hemisphere populations have no problem with this date because it is winter for them. It all makes perfect sense, but re-read strike one!

▶ **Strike three:** The millions of miles of difference between closest and farthest approach don't produce a large enough difference in the seasons for high-latitude polar locations. The seasons near the equator have nearly uniform temperature throughout the year, which is much less variation than this difference in distance from the Sun would cause. All locations on Earth would have to share the same variation in temperature if distance changes caused the seasons.

The seasons have to be caused by something else, but what? Earth is tilted in comparison to its orbit around the Sun. The direction of this tilt changes very, very slowly (24,000 years for a single wobble!), so over the course of many years, the orbital motion of Earth carries its nearly uniform direction of tilt through a cycle in which everyone on the planet has a season with the Sun higher in the sky and a season with the Sun lower in the sky. This stability of Earth's tilt and its impact on Sun angle drastically changes the amount of solar energy absorbed by the surface of Earth and thus local temperatures. For locations other than the equator, this effect is enhanced by the correlation of longer hours of daylight with higher Sun angles and shorter days with lower Sun angles (Figure 7-17).

So what if Earth's orbit was more elliptical? If the orbit of a planet produces a large enough change in distance, then this could produce a larger effect on surface temperatures than that due to the tilt of the planet in its orbit. For planets with no tilt, distance from the Sun is the only cause of the seasonal changes in temperature.

> **Precession:**
> the change in alignment between the orbital plane and the tilt of an object's rotational axis over time due to torque. The same term applies to the wobble of a toy top.

Figure 7-17 On closest to the Sun day, note that Earth experiences two seasons at once, an impossibility if distance caused the seasons. A northern hemisphere observer experiences a lower-than-average noontime Sun and shorter days, while at the same instant in time a southern hemisphere observer experiences a higher noontime Sun and longer days. This change in solar intensity and duration is the primary cause of Earth's seasons.

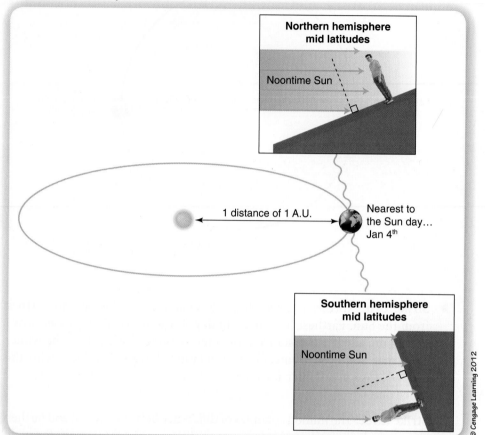

THE HUMAN FACTOR: COMPLICATING SPACE EXPLORATION

On December 19, 1972, the last humans returned to Earth from their journey to the Moon. When *Apollo 17* landed, it represented the last time a human was to travel beyond LEO. For more than 38 years, humans have traveled no farther into outer space than a typical commuter flight from Chicago to New York. All of our exploration of the solar system has been from the surface of Earth or carried out by robotic explorers in the form of orbiters, landers, and probes flown to Mars, Saturn, Titan, Pluto and beyond. But why have humans stepped back from the exploration of space with humans onboard?

As soon as a human is on the vehicle, we need to engineer solutions for pressurization, oxygen supply and CO_2 removal, food and water supplies, waste management, fitness, and psychological wellness. These tasks dramatically increase the sophistication of the vehicles required and increase the risk of catastrophic failure.

And yet we stand poised to send numerous missions to the Moon and establish a permanent outpost on its surface. First landers and orbiters will map out the surface and subsurface in greater detail, and then humans will return to the Moon's surface. The *Constellation* missions represent an opportunity for humans to pursue the dream of living among the heavens for very long periods of time. But it is not simply this romantic notion that drives us to explore. We have learned from long experience that the quest to explore leads us to innovation in science, technology, engineering, and products that change our daily lives. The combination of the two drives make it impossible for us to remain forever tethered to Earth.

THE RISK OF IMPACT: NEAR EARTH ASTEROIDS AND COMETS

Since hazards from asteroids and comets must apply to inhabited planets all over the galaxy, if there are such, intelligent beings everywhere will have to unify their home worlds politically, leave their planets. And move small nearby worlds around. Their eventual choice, as ours, is spaceflight or extinction.

Carl Sagan, *Pale Blue Dot*, 1994

The solar system contains many different objects, including the Sun, the planets, asteroids, and comets. Over the course of a human lifetime, we are likely to experience them directly through the cycles of day versus night and the seasons or the occasional "shooting star" visible as a small object burns up due to friction as it enters Earth's atmosphere at high velocity (Figure 7-18). We know from our personal experience that these collision events are common and harmless. However, as a species that communicates and writes down its history and with the tools of science, we are capable of asking "What are the odds of a really big object striking Earth? Wouldn't that be a little more devastating?"

NASA and other agencies have implemented observational programs using large automated telescopes to survey and catalog all of the objects in the solar system that are large enough to cause a significant change on Earth if we collide with them and that have orbits that would label them as a potential Near-Earth Object (NEO). This category implies that if the orbits of Earth and the object come close enough to each other in the future that a collision is at least possible. The goal of the survey is to find and identify every NEO of significant size and to determine and refine the orbital elements associated with them so that future impacts can be predicted. If we know their size and likelihood of impact, we can consider the devastation that they would cause and the investment that is worthwhile to try to prevent the impact (Figure 7-19).

Based on our understanding of orbital mechanics, we have come up with a small number of methods for changing Earth's interaction with these future impactors. The three primary methods include: shatter the object into a large number of very small objects, gently nudge the object over a long period of time to modify its orbit, or generate a massive push on the object over a short period of time to drastically alter its orbital path.

The surveys that have been completed through 2010 continue to find more and more objects that have orbits that will bring them near Earth in the future; however, the number of large objects being discovered has leveled off, implying that we have identified almost all of the large objects likely to cause harm to us (Figure 7-20). NASA continues to observe NEO and refine our data because the orbits of objects in space can change when they interact with large gas giants such as Jupiter and Saturn.

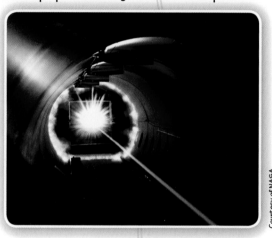

Figure 7-18 Energy flash from an impact at 17,000 miles per hour. The blue streak is due to the heating of the air in the chamber and is similar to the "shooting star" effect observed by your eye when small objects burn up upon entering Earth's atmosphere.

Courtesy of NASA.

NEO (Near Earth Object): an asteroid or comet with an orbit that intersects or nearly intersects with that of planet Earth in terms of alignment in both space and time; thus, an object with a significant likelihood of impacting Earth.

Figure 7-19 Energy released by an impacting object and the average time between impacts of this size.

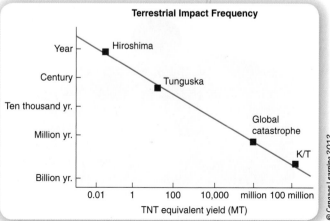

Terrestrial Impact Frequency

© Cengage Learning 2012

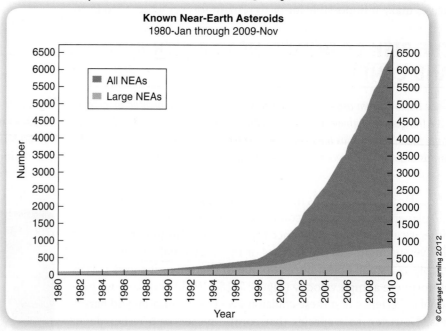

© Cengage Learning 2012

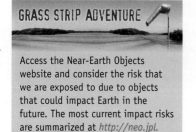

GRASS STRIP ADVENTURE

Access the Near-Earth Objects website and consider the risk that we are exposed to due to objects that could impact Earth in the future. The most current impact risks are summarized at *http://neo.jpl.nasa.gov/risk/*. Do you think it is worth our money, time, and effort to prevent future impacts?

NEOs are potentially dangerous to life on Earth due to the energy a collision would release. To understand the risk to Earth's inhabitants, we would need to assess the immediate impact of earthquakes, heat and debris blasts, and tsunami waves, as well as the longer-term impact on the atmosphere, global climate, and water quality (Figures 7-21 and 7-22). As a species, we possess the technological capabilities to discover potential impact objects, fully describe their orbital characteristics, and intervene to alter their path through space to prevent an impact from occurring. But should we use this technology? Like any insurance policy, it is all about evaluating the likelihood of an impact and considering the outcome should this unlikely event occur.

Figure 7-21 This illustration suggests what an impact by a 500 km diameter object might look like during the initial collision.

© Cengage Learning 2012

Figure 7-22 Artist's conception of the tsunami-like waves that would inundate coastal areas in the aftermath of a large object colliding with the Earth.

Courtesy of NASA

Careers in Aerospace Engineering

MAN ON A MISSION

Leland Melvin

Even when he was trying out for the Detroit Lions, Leland Melvin knew he needed a back-up plan. The wide receiver for the University of Richmond football team had majored in chemistry and had always considered himself more a student-athlete than an athlete-student. But the Lions had drafted him in the 11th round, and this was his shot at playing pro ball.

Unfortunately—or was it fortunate?—Melvin pulled a hamstring during training camp and was dropped from the team roster. The following spring, the same thing happened: He went down for a pass, pulled the hamstring again, and got dropped from the roster, this time by the Dallas Cowboys.

Leland Melvin's pro football career was over before it began.

So, he went with Plan B, which eventually took him not to the Super Bowl but to outer space. After working for NASA's Langley Research Center, Melvin was selected for the astronaut corps in 1998. A decade of preparation later, he was aboard the space shuttle Atlantis, speeding toward the International Space Station. And a year after that, he was on board again. Altogether, Melvin has logged over 565 hours in space.

On the Job

Many of those hours were spent working a robotic arm that, on the first mission, attached a new laboratory to the space station and, on the second mission, helped transfer spare parts from the shuttle. "I always tell kids it's like using a joystick," says Melvin, who's done a lot of work for NASA in the areas of education and outreach. "If I can do it, they can do it."

Inspiration

Both of Melvin's parents were teachers, but it was during summer vacations when his love for engineering would kick in. "We would go on these camping trips, and along the way I'd see tractor-trailers and trucks and trains," he says. "I was always fascinated by them. How were they made? Who put them together? Here were these vehicles that were traveling to far-off destinations."

Closer to home, Melvin fondly remembers the chemistry set his mother bought for him. "I made a little bomb in the living room," he says, chuckling, "but what I especially remember is taking these two clear liquids, mixing them together, and watching them transform into a deep-purple liquid."

Education

Not long after that, Melvin was majoring in chemistry at the University of Richmond, where he also had a football scholarship. But there was never any doubt which came first. "I asked the head coach what to do when there was a conflict," Melvin says, "and he said, 'You will go to class instead of practice.' So, on Tuesdays and Wednesdays, I went to chem lab."

An MS in materials science engineering from the University of Virginia helped Melvin land a job at NASA's Langley Research Center, where he led a project to turn optic fiber into sensors that could be used [Melvin: Can you fill out the rest of this sentence? Used to do what?] on space vehicles. "It was really fun," he says, "because we started from scratch, set up a lab and made our own optic fibers. And on the day we made our first sensor, I got the call to come and be an astronaut."

Advice

Want to be an astronaut yourself? Here's what Melvin suggests: Study hard. "You know how teachers will assign the odd-numbered exercises?" he asks. "Well, do the odd ones and the even ones, because the even-numbered ones will probably be on the test. Do a hundred extra problems. Practice makes perfect."

S122-E-008896 (15 Feb. 2008)—Astronaut Leland Melvin, STS-122 mission specialist, lends his intravehicular support to the two STS-122 mission specialists assigned to the mission's final spacewalk to perform work on the International Space Station. Equipped with their extravehicular mobility units (EMU) and other gear and just about ready to egress the station and begin the day's external tasks are astronauts Stanley Love (left) and Rex Walheim.

SUMMARY

- Orbits carry an object around another object due to the gravitational attraction between the two objects.

- Orbits follow three basic laws that were first identified by Johannes Kepler. Kepler's first law requires that an object orbit in an elliptical path with a circle being a special case of an ellipse. His second law demands that the object moves faster when it gets closer and moves slower when it gets farther from the object it is orbiting. This identifies two unique locations in the orbit referred to as aphelion and perihelion for orbits around the Sun and apogee and perigee for orbits around the Earth. Kepler's third law relates the square of the time it takes to complete one orbit to the cube of the size of the orbit itself.

- Gravity is the force controlling the shape of an orbit. If an orbiting object has just the right speed, it will remain in the equilibrium condition of a uniformly circular orbit in which it remains at a constant distance and speed in relationship to the object that it is orbiting.

- The orbit of an object can be fully described by a small number of characteristics called orbital elements. These elements change over time due to the complex interaction of the gravity from multiple objects and forces created through collisions or human actions such as rocket burns.

- Transferring from orbiting one object to orbiting another requires an engine burn to change the orbital elements so that one part of the orbit intercepts the target object that you seek to orbit. Most transfers require a second burn upon arrival to once again change the object's orbital elements to produce a stable orbit around the new object.

- The gravitational force from a planet can be used to significantly speed up and redirect an object in its journey through a gravitational slingshot maneuver.

- There is a common misconception that Earth's seasons are directly caused by our changing distance from the Sun. This is obviously false because one distance cannot produce two seasons at the same time. The real cause for experiencing different seasons simultaneously at various Earth locations is that its orbit carries its nearly uniform tilt (compared to the orbital plane) around the Sun, giving one hemisphere a period of higher noontime Sun and longer days followed by a period of lower Sun and shorter days.

- Human exploration of space adds complexity to the requirements for the space vehicle but simplifies the task of carrying out the science of exploration.

- The greatest threat to life on planet Earth is most likely the impact of a large asteroid or comet that has an orbit that intersects Earth's orbit at the same instant in time. Researchers and engineers are identifying potential impact objects, describing their orbits fully, and making plans for altering their orbits to prevent an impact from occurring.

BRING IT HOME

1. Make a loop out of a string that is about 8 inches long, and then use the string, a pencil, and two thumb tacks to draw a number of ellipses with a single focal point (1 tack) compared to two focal points (2 tacks) with the focal points placed at different distances from each other. What happens to the shape as the focal points move apart?

2. According to Kepler's second law of planetary motion, what must be true about the orbital velocity of an object as it moves from perihelion to aphelion? How about from aphelion to perihelion?

3. Considering Kepler's third law of planetary motion, what has to happen to the orbital period of the Space Shuttle if it moves from an LEO up to the orbit of the International Space Station for docking?

4. Describe the series of rocket engine burns that are required to transfer from an LEO to an orbit around the Moon, then onward to Mars, and then back home again. You can keep it simple by just describing the burn as either a boost (in the existing direction of flight) or a retro (opposite the direction of flight) firing.

EXTRA MILE

1. Compile a list of products that were invented to support NASA missions that have become commonly used products during the past 40 years.

2. Calculate the orbital period for a comet that travels through space at three times Earth's orbital radius around the Sun.

3. Rank this list of the materials and equipment that are available for inclusion in your spacecraft's survival kit. Prioritize the list so that the most important material/equipment is listed first.

   ■ **Food concentrate:** Dehydrated food, just add water and mix.

   ■ **Space blanket:** A thin, shiny blanket made out of highly reflective Mylar.

   ■ **First-aid kit:** Standard medical supplies such as bandages and antiseptics.

   ■ **Matches:** A box of standard fireplace matches.

   ■ **Space suit patch kit:** Contains adhesive and impermeable fabric patches.

   ■ **Compass:** Device that indicates the direction of a magnetic field.

   ■ **Life raft:** A self-inflating ring with a flexible rubber bottom.

   ■ **Tarp:** A 20 ft by 20 ft square of heavy-duty reinforced plastic tarp.

   ■ **Map:** Indicating the location of places on the Moon.

   ■ **Radios:** Two radio transceivers with batteries.

   ■ **Tanks of oxygen:** Large pressurized tanks of pure oxygen.

   ■ **LED flashlights:** Flashlights that include batteries.

   ■ **Signal mirrors:** Reflective mirrors.

   ■ **Rope:** 100 ft of nylon rope.

   ■ **Tanks of water:** Large, flexible canisters of pure water.

4. Investigate how new asteroids and comets are discovered and named. What equipment is needed, and how is it used?

5. Seek out a local astronomy club to find a local expert on the subject of asteroids and comets to come and talk to your class.

CHAPTER 8
Aerospace Physiology

GPS
DELUXE

| START LOCATION | DISTANCE | END LOCATION |

Menu

Before You Begin

Think about these questions as you consider the concepts in this chapter:

1. What is physiology, and why is it of concern to an aerospace engineer?

2. How does each body system function during our normal daily experiences?

3. How is oxygen used by our bodies, and what happens when it isn't available?

4. How do aggressive maneuvers, or "pulling Gs," affect the pilot and crew?

5. What happens when our bodies become too warm or cold?

6. How do we "see" our surroundings?

7. What is spatial disorientation, and how is it overcome?

8. How is the human body affected by long-duration space flight?

9. What training and techniques are used to overcome the limits of our physiology?

10. What role has technology played in allowing humans to expand their ability to fly?

Human physiology is the study of the chemical and physical characteristics and functions of a normal, healthy individual. Physiologists seek to understand how our cells function as groups to carry out basic life functions.

The human body is not naturally adapted to operate under the conditions typical of the aerospace environment. Our lack of adaptation takes the traits that ensure our survival on a daily basis and turns them into potentially fatal weaknesses in the aerospace environment. In almost every case, the normal human structures and functions you studied in biology class are revealed to fail under the additional stress of this environment. Even so, our engineered solutions have allowed us to sustain the human body almost anywhere on Earth and far out into space.

As engineers, our goal is to understand the human body's limits and to consider how they might be overcome through the development of equipment or procedures that enhance the human body or limit its exposure to dangerous situations.

Humans have always sought to explore the extreme environments found above and beyond Earth's surface. This chapter introduces some of the physiological principles that aerospace engineers must use to safeguard our well-being as we venture far from the comfort of the world to which we are so well adapted.

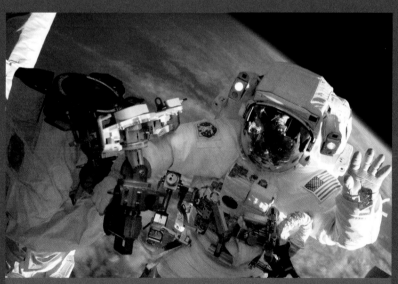

Astronaut Clay Anderson waves to the camera during a 7-hour, 41-minute spacewalk outside the International Space Station. Anderson and cosmonaut Fyodor N. Yurchikhin installed a television camera stanchion, reconfigured a power supply for an antenna assembly, and performed several construction tasks.

Courtesy of NASA

THE BODY IN FLIGHT

3

Understanding how human physiological systems work is key to understanding how they will be affected by stresses created during aerospace flight. Engineers will work closely with human factor experts when designing aerospace craft, but it helps to have a solid knowledge of the anatomy behind human physiology.

Oxygen

As you can see in Figure 8-1, Earth's **atmosphere** is actually quite thin compared to the overall size of Earth. The predominant gasses that make up the atmosphere are nitrogen and oxygen (see Table 8-1). Every cell in the human body requires oxygen to release the energy that allows the cell to carry out its assigned function. This reaction is known as **cellular respiration**.

During respiration, the cell uses enzymes to break down nutrients and manufacture adenosine triphosphate (ATP). The ATP molecule is a universal energy source for cells. The waste product of the cellular respiration reaction is carbon dioxide.

Oxygen in the Human Body

Breathing is a continuous cycle of inhalation and exhalation. Air moves into and out of our lungs due to a pressure difference between the inside of the lung and the air around us. To start an inhalation, the strong muscles of the diaphragm contract pulling the domed structure downward. This expands the cavity in which the lungs are contained, which reduces the pressure surrounding the lungs. Normal air pressure around our body pushes air through our mouth and nose, down the throat and trachea into the lungs causing them to expand until the pressure is equalized. This brings fresh air into the lungs (see Figure 8-2). Air is exhaled when the diaphragm relaxes. This reduces the lung cavity's size and increases the pressure surrounding the lungs. Combined with the elastically expanded ribs, this elevates pressure inside the lungs above surrounding pressure, and the air flows out of the lungs.

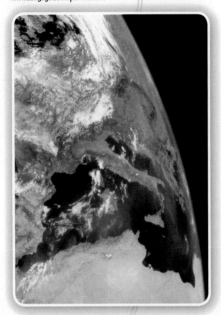

Figure 8-1 Earth's atmosphere is very thin compared to the size of the planet. In this image, it is the pale blue halo seen at the edge of the Earth image.

From AHRENS. *Meteorology Today: An Introduction to Weather, Climate, and the Environment,* 9th edition. © 2009 Brooks/Cole, a part of Cengage Learning. Reproduced with permission. www.cengage.com./permissions

Table 8-1 Composition of the atmosphere near Earth's surface.

From AHRENS. *Meteorology Today: An Introduction to Weather, Climate, and the Environment,* 9th edition. © 2009 Brooks/Cole, a part of Cengage Learning. Reproduced with permission. www.cengage.com./permissions

Permanent Gases			Variable Gases			
Gas	Symbol	Percent (by Volume) Dry Air	Gas (and Particles)	Symbol	Percent (by Volume)	Parts per Million (ppm)*
Nitrogen	N_2	78.08	Water vapor	H_2O	0 to 4	
Oxygen	O_2	20.95	Carbon dioxide	CO_2	0.038	385*
Argon	Ar	0.93	Methane	CH_4	0.00017	1.7
Neon	Ne	0.0018	Nitrous oxide	N_2O	0.00003	0.3
Helium	He	0.0005	Ozone	O_3	0.000004	0.04†
Hydrogen	H_2	0.00006	Particles (dust, soot, etc.)		0.000001	0.01–0.15
Xenon	Xe	0.000009	Chlorofluorocarbons (CFCs)		0.00000002	0.0002

*For CO_2, 385 parts per million means that out of every million air molecules, 385 are CO_2 molecules.
†Stratospheric values at altitudes between 11 km and 50 km are about 5 to 12 ppm.

Figure 8-2 Location of the diaphragm and changes in the thoracic cavity during inhalation and exhalation with X-rays.

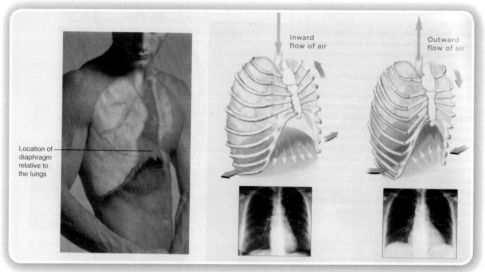

Lungs are structured around a system of tubes that start in the throat. The bronchus branches into two bronchi, which branch into multiple bronchioles each ending in a clump of small airspaces called alveoli. Each clump of alveoli is surrounded by capillary blood vessels. At this interface between the air in the alveoli and blood cells, oxygen, carbon dioxide, and other gasses are exchanged (see Figure 8-3). Spread out uniformly, the surface area between the lungs and the bloodstream is

Figure 8-3 Two examples of diffusion: (a) A drop of dye enters a bowl of water. (b) The same thing happens with the water molecules. Here, red dye and yellow dye are added to the same bowl. Each substance will move down its own concentration gradient.

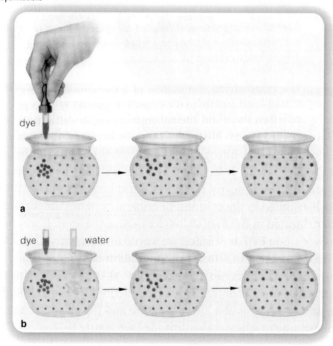

Figure 8-4 Components of the human respiratory system and their functions. Also shown are the diaphragm and other structures with secondary roles in respiration.

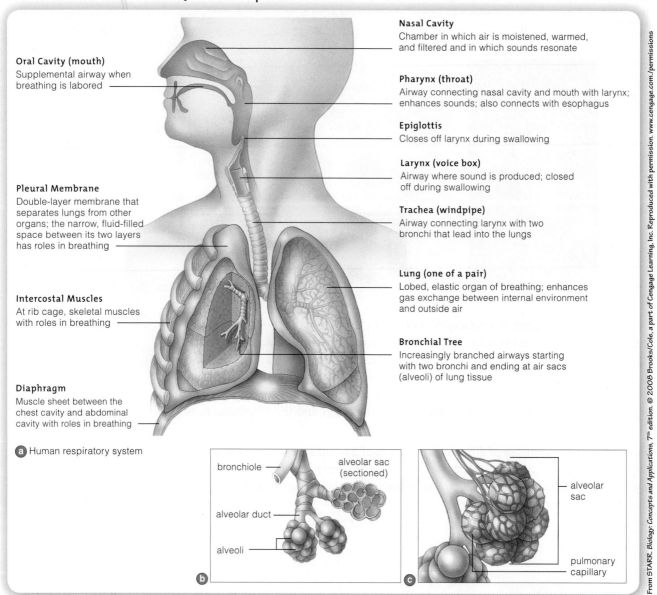

Oral Cavity (mouth)
Supplemental airway when breathing is labored

Pleural Membrane
Double-layer membrane that separates lungs from other organs; the narrow, fluid-filled space between its two layers has roles in breathing

Intercostal Muscles
At rib cage, skeletal muscles with roles in breathing

Diaphragm
Muscle sheet between the chest cavity and abdominal cavity with roles in breathing

ⓐ Human respiratory system

Nasal Cavity
Chamber in which air is moistened, warmed, and filtered and in which sounds resonate

Pharynx (throat)
Airway connecting nasal cavity and mouth with larynx; enhances sounds; also connects with esophagus

Epiglottis
Closes off larynx during swallowing

Larynx (voice box)
Airway where sound is produced; closed off during swallowing

Trachea (windpipe)
Airway connecting larynx with two bronchi that lead into the lungs

Lung (one of a pair)
Lobed, elastic organ of breathing; enhances gas exchange between internal environment and outside air

Bronchial Tree
Increasingly branched airways starting with two bronchi and ending at air sacs (alveoli) of lung tissue

bronchiole

alveolar sac (sectioned)

alveolar duct

alveoli

ⓑ

alveolar sac

pulmonary capillary

ⓒ

From STARR. Biology: Concepts and Applications, 7th edition. © 2008 Brooks/Cole, a part of Cengage Learning, Inc. Reproduced with permission. www.cengage.com./permissions

approximately equal to a tennis court. Gasses pass from the air to the bloodstream and vice versa by diffusion based on the number of actual molecules on each side of the alveoli surface (see Figure 8-4).

Normal air contains 78% nitrogen and 21% oxygen. Because the atmosphere is old and well mixed, these percentages remain constant nearly everywhere on Earth. Earth's atmosphere is nearly 100 miles thick; however, the air molecules are not uniformly distributed in the column. Because air can be compressed, half of all air molecules are located in the first 18,000 feet above sea level. If we could weigh a column of air over a 1-in by 1-in surface, we would find that at sea level, the column would weigh approximately 14.7 lbs. If the mountain climber in Figure 8-5 was at 18,000 feet, or a pilot ascending in an aircraft was at the same height, the column above the scale would weight only 7.35 lbs (see Figure 8-6). So although the composition is the same, the air is at a lower pressure and the air molecules spread out to fill twice the volume of space at this altitude (see Figure 8-7).

Figure 8-5 For a mountain climber or pilot above 18,000 ft, atmospheric pressure is 500 mb or 7.35 lbs/in².

From STARR. Biology: Concepts and Applications, 7ᵗʰ edition. © 2008 Brooks/Cole, a part of Cengage Learning, Inc. Reproduced with permission. www.cengage.com./permissions

Figure 8-6 Atmospheric pressure decreases rapidly with altitude. Climbing to an altitude of only 5.5 Km (18,000 ft), where the pressure is 500 mb or 7.35 lbs/in², would put you above one-half of the atmosphere's molecules.

From AHRENS. Meteorology Today: An Introduction to Weather, Climate, and the Environment, 9ᵗʰ edition. © 2009 Brooks/Cole, a part of Cengage Learning. Reproduced with permission. www.cengage.com./permissions

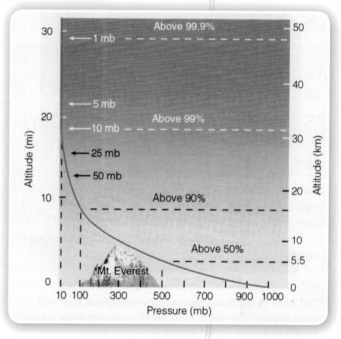

Figure 8-7 shows the decrease of molecules in the atmosphere as altitude increases. As an aircraft ascends, the concentration of oxygen remains 21% while the actual number of molecules in a "lung full" of air continuously decreases. This means the higher a person climbs in altitude, the more slowly oxygen enters the blood-stream. At some altitude, no oxygen will enter the bloodstream, and oxygen will actually flow out of the bloodstream with each breath at even higher altitudes. The reduction in air pressure leads to an outcome referred to as **time of useful consciousness** (see Table 8-2).

A pilot whose blood is saturated with oxygen will have full mental capabilities available to deal with normal and emergency flight procedures. If the pilot is exposed to the reduced pressure at altitude, the body's ability to replenish oxygen in the bloodstream will lag behind the demand for oxygen from the bloodstream by cells carrying out respiration. At some point, the blood oxygen level will fall low enough that the brain will start to shut down. The pilot might be conscious but will be incapable of taking the correct actions required to safely fly the plane. The higher the aircraft is flying, the less time it takes to reach this point.

Figure 8-7 Both air pressure and air density decrease with increasing altitude. The weight of all of the air molecules above Earth's surface produces a sea level pressure of 14.7 lbs/in² (PSI).

From AHRENS. Meteorology Today: An Introduction to Weather, Climate, and the Environment, 9ᵗʰ edition. © 2009 Brooks/Cole, a part of Cengage Learning. Reproduced with permission. www.cengage.com./permissions

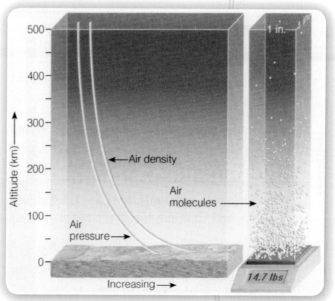

Table 8-2 Time of useful consciousness reflects how quickly the brain becomes hypoxic at different altitudes. After the indicated time has passed, the pilot in command cannot be relied upon to make the decisions required to continue the flight safely.

Source: Pilot's Handbook of Aeronautical Knowledge, FAA Manual 8083-25A, www.faa.gov

Altitude	Time of Useful Consciousness
45,000 feet MSL	9 to 15 seconds
40,000 feet MSL	15 to 20 seconds
35,000 feet MSL	30 to 60 seconds
30,000 feet MSL	1 to 2 minutes
28,000 feet MSL	2½ to 3 minutes
25,000 feet MSL	3 to 5 minutes
22,000 feet MSL	5 to 10 minutes
20,000 feet MSL	30 minutes or more

Point of Interest

Hollywood and Diving Airplanes

Watch enough movies about airplanes and you are bound to run into a scene where the aircraft loses pressure, the oxygen masks fall from the overhead rails, and the airplane begins to plunge to the surface (see Figure 8-8). Many people assume that this means airplanes can't fly if they lose cabin pressure, but what you are seeing is a well-trained pilot taking appropriate action.

Pilots know that after they lose pressure in the cabin, the clock is ticking in terms of time of useful consciousness. As soon as pressure is lost, the pilot puts on his or her own oxygen mask and then begins a maximum safe rate of descent dive to bring the aircraft below 10,000 feet where it is safe to fly for extended periods of time. After descending below this altitude, the pilot will pull out of the dive and continue the flight to the nearest airport.

Figure 8-8 Movies always show the aircraft diving after the cabin loses pressure. Pilot's are trained to perform a maximum safe speed dive to an altitude below where supplemental oxygen is required as airliners carry a limited supply of emergency oxygen onboard. The aircraft isn't going to crash, it just needs to lose altitude quickly.

© Cengage Learning 2012

Hypoxia

Hypoxia is a reduction in the amount of oxygen that is available for cellular respiration and is a result of a breakdown in the human body's ability to gather, absorb, and deliver the oxygen to every cell. The most common causes for hypoxia are flight at altitude, maneuvering flight, and carbon monoxide poisoning. Signs of hypoxia include rapid breathing, blue-tinged lips and fingernails, loss of coordination, lethargy, and poor judgment.

Oxygen Systems and Regulations

The FAA requires all aircraft operating at high altitude to provide oxygen to the flight crew and passengers. Below 10,000 ft, pilots can fly indefinitely without supplemental oxygen; however, it is strongly recommended for flights above 5,000 ft at night. Above 12,000 ft, the flight crew is required to have supplemental oxygen on longer flights; however, there is no requirement to provide any oxygen to passengers until the flight climbs above an altitude of 15,000 MSL (mean sea level).

For large commercial aircraft, the simplest solution is to pressurize the entire cabin to an equivalent altitude that is below 12,000 ft of altitude. Should the cabin lose pressure, the aircraft has backup oxygen masks that drop from the overhead rail to provide pure oxygen to every passenger for the few minutes required to complete an emergency descent.

Another solution is to provide oxygen directly to the person via a mask. This eliminates the need to pressurize the entire cabin to the lower altitude. However, this method does have a limit. At extremely low pressures, the boiling point for water approaches body temperature. Many military high-altitude flights are thus conducted in partial or full pressure suits that contain the pilot's entire body in a bubble of air at an elevated pressure.

Fun Fact

Cooking at Altitude

Go to the cupboards and look at the cooking instructions for baked goods such as cakes, cupcakes, and brownies. You'll notice a pattern in that almost all baked goods include high-altitude cooking directions. Think before you leap . . . how should the directions be modified to account for high-altitude cooking? Check the labels; were you correct?

In spacecraft, we have a unique need in that every pound brought to orbit costs many thousands of dollars. Remember that 21% of air is oxygen, and 14.7 lbs per square inch (psi) is normal atmospheric pressure. Note that 21% of 14.7 is a little more than 3 psi, which means that astronauts can breathe normally at a pressure of only 3 psi if the atmosphere is pure oxygen. By using pure oxygen environments, we eliminate the need to haul 79% of the gasses to orbit.

How Much Weight for the Air?

We normally don't think of the air around us weighing anything because the pressure is equal all around us. When you take that air in a container (the Space Shuttle, for example) and place it in the near vacuum of space, the weight of the air becomes measurable. This is important when we consider that it can cost as much as $10,000 per pound of material brought to orbit.

Problem

Given that the volume of the crew cabin of the Space Shuttle is approximately 2,625 ft³, and standard air is about 1.2 oz/ft³. How much weight is saved by only using 3 psi of pure oxygen in the Space Shuttle compared to keeping the shuttle pressurized with standard air at sea level (14.7 psi)? (Remember approximately 20% of the atmosphere is oxygen, so 20% of 14.7 is roughly 3 psi.)

Solution

Weight of Air on shuttle at normal air pressure 1.2 oz/ft³ * 2625 ft³ = 3150 oz

3150 oz/16 oz/1 lb = 197 lbs

Weight of Air using only pure oxygen at 3 psi = 0.20 × 1.2 oz = 0.25 oz/ft³

0.25 oz/ft³ * 2625 ft³ = 656 oz

656 oz/16 oz/1 lb = 41 lbs

Difference of 197 lbs − 41 lbs = 156 lbs saved just by changing to pure oxygen!

Circulation and Blood Pressure

To pass oxygen to the various physiological systems, the oxygenated blood must get circulated through the body. This is the primary function of the circulatory system. The heart is the pump. When it contracts, it creates high pressure. The valves in the heart ensure that the blood can only flow in one direction. Every time the heart contracts, the blood is sent out into both the lungs and the body. The loop that passes through the lungs exchanges carbon dioxide for oxygen and returns the oxygenated blood back to the heart. The systemic loop carries oxygenated blood out into the body where cells absorb the oxygen for use in respiration and return carbon dioxide to the blood. Both loops are closed systems where gasses are added and removed from the bloodstream while the blood is in motion, as Figure 8-9 illustrates.

Most of the oxygen in your blood is consumed by one organ, the brain. The heart has to create enough blood pressure to push the blood rapidly upward 15 inches from the heart to the brain. For a healthy heart, the forcefulness of each contraction is easily capable of carrying out this lifting task (see Figure 8-10).

G-Forces

Imagine how much strength you would need to lift a brick from the seat of your desk to the desktop. This is not a very large force. But what if you needed to lift two bricks at a time? How about eight bricks at a time? How many bricks would it take to make the task so hard that you don't have the strength to lift the stack of bricks at all?

Figure 8-9 *The cardiac cycle.*

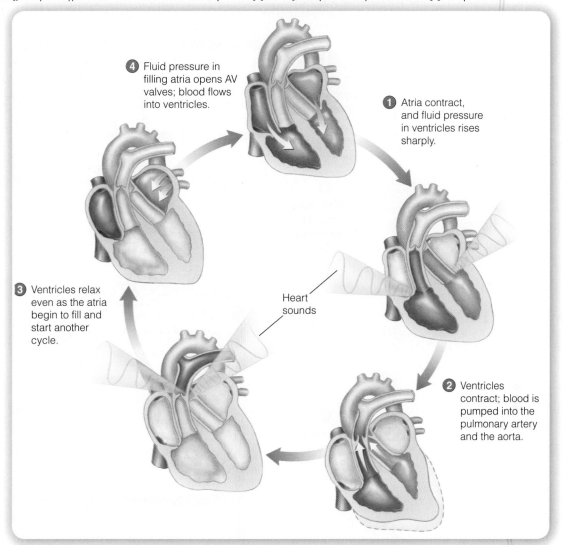

4 Fluid pressure in filling atria opens AV valves; blood flows into ventricles.

1 Atria contract, and fluid pressure in ventricles rises sharply.

3 Ventricles relax even as the atria begin to fill and start another cycle.

Heart sounds

2 Ventricles contract; blood is pumped into the pulmonary artery and the aorta.

The blood in your body has weight due to the pull of gravity. More accurately, you feel the force of weight due to the acceleration that gravity exerts on your body. So what happens when we maneuver the aircraft by rolling it up into a steeply banked turn? We create accelerations that our bodies perceive as additional gravity forces or **G-forces** (see Figure 8-11). This means that our heart has to push twice as hard to lift the blood to the brain. Normal flight maneuvers will produce G-forces as large as 2 Gs and are tolerable by almost everyone (see Figure 8-12).

Roll into a steeper turn or pull up quickly and pilots can pull 4, 6, 9, or even more Gs (see Figure 8-13). At some point, the heart cannot create sufficient pressure to lift the blood into the brain, and circulation fails. The pilot and passengers will suffer stagnation hypoxia. The oxygen reserve in brain and retinal tissue in the eye runs out. Within seconds, the pilot will transition through the loss of color vision (grayout), tunnel vision, and blackout as the pilot experiences **G-Induced Loss of Consciousness (G-LOC)**. The pilot's brain will remain offline until the G-force is reduced and circulation returns, followed by minutes of reduced higher-order thinking abilities. If oxygen does not return soon enough, the brain remains anoxic (without oxygen) long enough that death occurs due to the crash that results from the aircraft flying without an able-bodied pilot.

G-forces:

the perceived multiplier for weight. Maneuvering flight can change the multiplier to a negative value, zero (perceived weightlessness), or any positive value.

Figure 8-10 The (a) pulmonary and (b) systemic circuits for blood flow in the cardiovascular system.

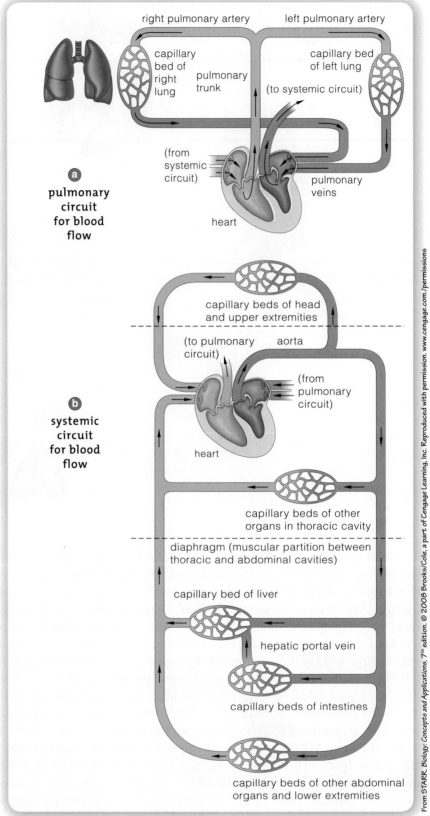

right pulmonary artery

left pulmonary artery

capillary bed of right lung

pulmonary trunk

capillary bed of left lung

(to systemic circuit)

(from systemic circuit)

pulmonary veins

heart

a

pulmonary circuit for blood flow

capillary beds of head and upper extremities

(to pulmonary circuit)

aorta

(from pulmonary circuit)

b

systemic circuit for blood flow

heart

capillary beds of other organs in thoracic cavity

diaphragm (muscular partition between thoracic and abdominal cavities)

capillary bed of liver

hepatic portal vein

capillary beds of intestines

capillary beds of other abdominal organs and lower extremities

Figure 8-11 The number of Gs that are experienced by the aircraft and humans on board is directly related to how quickly the aircraft is maneuvering. This chart relates the bank angle of a turn in which the aircraft is holding altitude to the G-force multiplier for their weight.

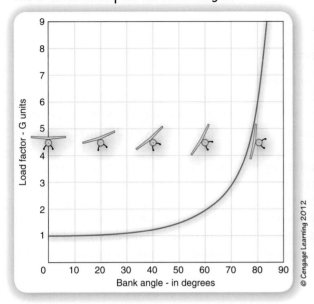

Figure 8-12 The brain and eyes consume a large portion of the oxygen in the bloodstream and are very sensitive to a loss of blood flow. The effective distance of these organs from the heart is equal to their true distance multiplied by the G-force as the blood flowing to them becomes heavier and more difficult to pump upward against gravity's pull.

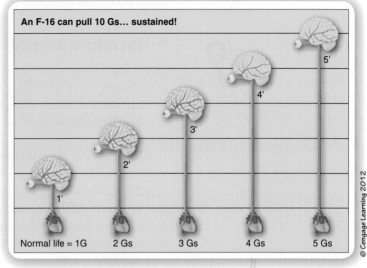

Figure 8-13 When pulling Gs during aerial combat maneuvers, it is not unusual for fight pilots to experience 9 to 11 times their normal body weight. This means that their heart has to work as if it were pumping blood an incredible 10 feet to reach the brain!

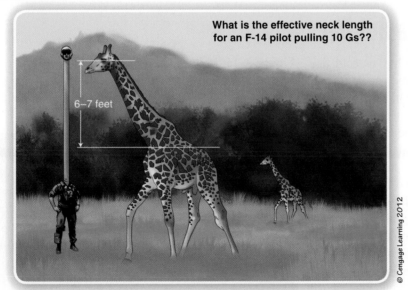

To prevent G-LOC and reduce its impact on the safety of flight, pilots are trained to perform a "grunt" that has them squeeze every muscle in their lower bodies. This can be enhanced by wearing a **g-suit** that contains air bladders that are hooked to the airplane and inflate nearly instantaneously when elevated G-forces are experienced. The g-suit and grunt work together to elevate the blood pressure in the lower extremities to prevent blood flow downward out of the brain. Modern aircraft can also be programmed to prevent the airplane from exceeding a maximum G-force and to pull the aircraft into a straight and climbing flight path if its computer thinks the aircraft is about to crash, and the pilot is not taking corrective actions. These systems allow fighter pilots to routinely fly at more than 9 Gs for short periods of time.

Temperature Regulation

The human body works hard to maintain a uniform body core temperature of approximately 98° F (37° C). Changing our core body temperature by even a small amount leads to major changes in our physiology. Reducing our body temperature is called hypothermia, while an increased temperature is called hyperthermia or heat stroke.

Reducing our body temperature by a few degrees causes shivering of muscles as they attempt to release more heat and goose bumps to make the hair on our bodies stand up to thicken the insulating boundary layer around our bodies. Drop another few degrees, and the shivering becomes extreme, the muscles begin to fail to contract, and we become uncoordinated as blood circulation shuts down, and the skin becomes pale or blue. Severe hypothermia produces a stupor, irrational thought, and ultimately death.

Point of Interest

Freezing to Death on a Hot Day

We are accustomed to defining a hot or cold day based on how it feels to us. However, the definition of hot is based on the speed of the molecules in the air. To quickly transfer energy from the air to our bodies by conduction requires billions of molecular collisions per second. At extreme altitudes, the average speed of the air molecules is high enough to be above 1000 °C, but there are so few molecules that astronauts wouldn't gain much energy at all by conduction from the extremely thin air surrounding them (see Figure 8-14). So how would they freeze? Recall that there are three ways to transfer energy: conduction, convection, and radiation. The astronauts would radiate away their body heat much quicker than it could be replaced from the air, so they could quickly freeze to death in the shadow of their spacecraft.

Figure 8-14 How can an astronaut survive when the "air" outside is 1000 °C?

From AHRENS. *Meteorology Today: An Introduction to Weather, Climate, and the Environment,* 9th edition. © 2009 Brooks/Cole, a part of Cengage Learning. Reproduced with permission. www.cengage.com./permissions

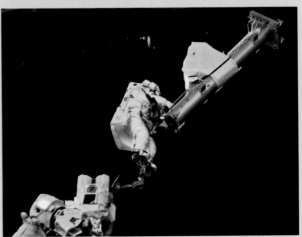

Figure 8-15 Layers of the atmosphere and their relationship to air temperature above Earth's surface. The red line indicates how the average temperature varies in each layer.

From AHRENS. *Meteorology Today: An Introduction to Weather, Climate, and the Environment*, 9th edition. © 2009 Brooks/Cole, a part of Cengage Learning. Reproduced with permission. www.cengage.com./permissions

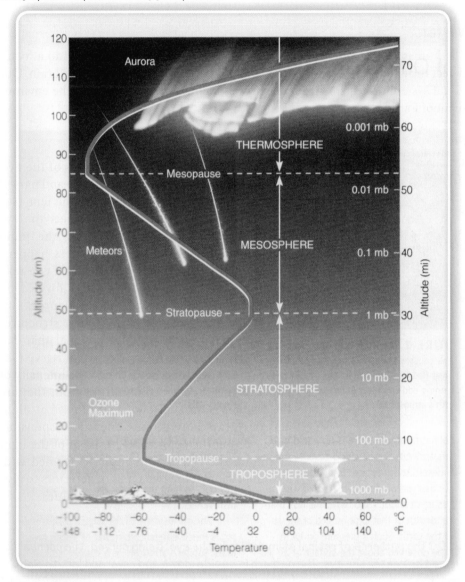

In the aerospace environment, the atmospheric temperature decreases by 3.5° F (2° C) for every thousand feet of altitude up to approximately 36,000 ft MSL (see Figure 8-15). From this point on, the temperature remains constant at nearly −70° F up to more than 60,000 ft MSL. This means that all aircraft operating at these high altitudes and pilots that bail out of the aircraft require a source of heat to survive. In outer space, the challenge is worse because the intense radiation from the Sun can raise the illuminated side of the spacesuit to hundreds of degrees above zero while the dark side of the suit cools to hundreds below zero.

Hyperthermia results when the body's normal cooling system of sweating is incapable of keeping up with the demands placed on it. Heat stroke is a rapid rise in temperature to more than 110° F and requires the immediate cooling of the body. Heat exhaustion results from dehydration and lack of salt in the body. Both conditions are very dangerous and can lead to organ failure and death. In the aerospace environment the primary concern is for astronauts as they can quickly elevate their core body temperature in a spacesuit when exerting themselves in direct sunlight.

Vision

The human eye is composed of components that gather light, bring it to a focus, detect the light's intensity and color, and then transmits the signal to the brain. Our visual system protects us from dangers and allows us to function in the environment of Earth's surface.

Figure 8-16 Structure of the eye.

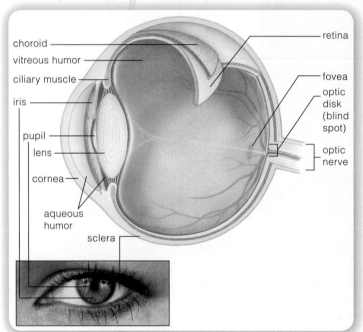

How We See

Figure 8-16 illustrates the major parts of the eye. Light is gathered at the front of the eye and passed through the pupil to be focused by the lens (see Figure 8-17). The iris serves to protect the eye from excessive light by constricting the pupil to reduce the light intake. This process happens very quickly, while reopening the pupil can take many minutes. The focused image is formed on the retina at the back of the eye (see Figure 8-18). The retina contains rods and cones, structures that detect light and produce a signal that can be passed to the brain via the optical nerve (see Figure 8-19). Rods are sensitive to the intensity of light but do not detect color information. They are the primary source of our ability to see at night and our grayscale vision. Cones are responsible for detecting the color characteristics of the light falling on the retina. Our retina is formed with a combination of the two types of detectors to give us both color and night-vision capabilities.

Figure 8-17 (a) How light can bend. (b) The patterns of retinal stimulation in the eye. Being curved, the cornea alters the trajectories of light rays as they enter the eye. The pattern is upside-down and reversed left to right, compared to the light source.

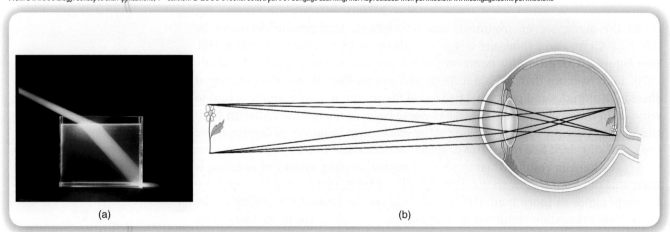

(a)　　　　　(b)

Figure 8-18 Light is focused on the retina by adjusting the lens (visual accommodation). Without stimulus, the eye relaxes to focus on a point about 6 ft away. This is important during flight in instrument conditions because it slows down the reaction speed of the pilot.

From STARR. *Biology: Concepts and Applications,* 7th edition. © 2008 Brooks/Cole, a part of Cengage Learning, Inc. Reproduced with permission. www.cengage.com./permissions

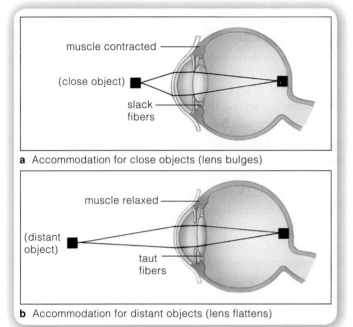

a Accommodation for close objects (lens bulges)

b Accommodation for distant objects (lens flattens)

Figure 8-19 Location of the foveal zone of high resolution, color vision, and the blind spot at the start of the optic nerve.

From STARR. *Biology: Concepts and Applications,* 7th edition. © 2008 Brooks/Cole, a part of Cengage Learning, Inc. Reproduced with permission. www.cengage.com./permissions

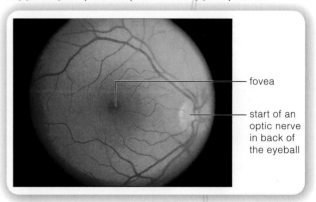

Cones are most concentrated at the **fovea**, which is the place where the lens forms its focused image (see Figure 8-20). Human foveal vision is very high resolution but over a very small angle, which means we have to look directly at something to see it clearly. Pick a letter on this page and stare right at it, and don't let

Figure 8-20 Cones provide our high resolution color vision and are concentrated near the fovea. Rods are triggered at much lower light levels and dominate our peripheral vision. Where the optical nerve enters the eye is a blind spot.

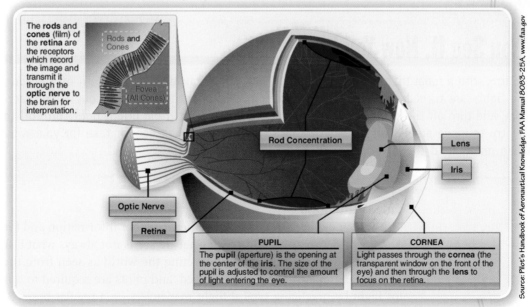

Source: Pilot's Handbook of Aeronautical Knowledge, FAA Manual 8083-25A, www.faa.gov

Figure 8-21 During the daytime, the bright light can trigger the cones, and we see with high resolution and color vision. At night, or in reduced lighting, the cones are not triggered. Lacking rods in the fovea, our central vision becomes a blind spot and averted visual techniques are required in order to see.

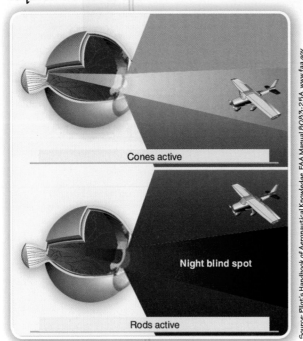

Source: Pilot's Handbook of Aeronautical Knowledge, FAA Manual 8083-25A, www.faa.gov

your eyes shift at all as you attempt to read the lines of text just above and below the letter you picked. Can you read one line away? How about two? Your foveal field-of-view is very narrow! Pilots are trained to use a visual scan of their environment to ensure that they can see and avoid other aircraft that are nearby.

Night Vision

As cones increase in concentration, rods must decrease in concentration, which means that the fovea is a night blind spot. You may have noticed that at night, you have trouble seeing very faint objects such as stars when you look directly at them. This is because the image of the object is falling on cones that require much more light to trigger and detect the light. However, if you look slightly off to the side of the object, its image will fall outside the fovea on the area where there are more rods than cones and lower levels of light are detected (see Figure 8-21). During the daytime, the pilot's scan for traffic includes short pauses ever few degrees to allow the foveal vision to see and avoid the traffic; however, at night this changes to a continuous scan to ensure that any traffic spends as little time as possible in the night blind spot.

The pupil contracts most strongly when the retina detects bright light that is white or blue in color. However, it barely responds at all to light that is pure red or green in color. People that have to work in the dark frequently use light sources that are red or green in color so that while the light is turned on, their pupils won't contract, and as soon as the light is turned back off, they can see in the dark. It is important for instrument panel and safety lighting in the cockpit to have suitable color and intensity to ensure that pilots maintain their night-vision capabilities.

Fun Fact

Now You See It, Now You Don't

Bring a mirror into a room that gets very, very dark when the lights are turned off. Go into the darkened room, and turn off the lights. Stay in the dark but don't try to move around for at least 10 minutes. When ready, hold the mirror a few inches in front of your eyes as you turn on the room lights. You can see how fast the pupils contract to protect your eyes from bright light. Turn off the lights again, how long does it take for your night vision to return?

Visual Illusions

The human eye and brain work together to gather information and form an image of the world around us. However, what we see is not always what is actually happening. To become skilled at interpreting the world as seen from the air, airport lighting and markings are standardized, and pilots are required to remain current for flight by practicing typical flight maneuvers on a regular basis. Because most of this practice takes place at the same few airports, the pilot can become over-comfortable with how the approach to landing looks for their particular home field.

When landing at an unfamiliar airport, pilots have to check the layout and dimensions of the runway as part of the preflight preparation (see Figure 8-22). This allows pilots to mentally rehearse what the approach is going to look like and avoid getting tricked by the illusions created by a runway that is narrower, wider, or more upsloped or downsloped than they are used to.

Figure 8-22 *Visual illusions during landing can cause the pilots to think they are flying a normal approach when the aircraft is actually too high or too low. Both conditions can be dangerous.*

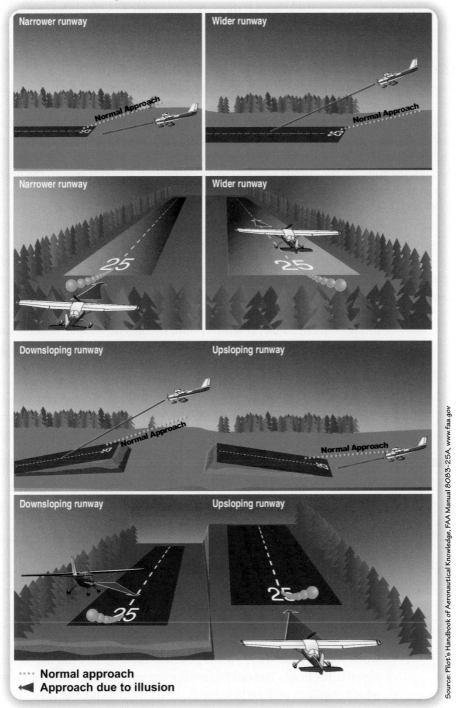

Source: Pilot's Handbook of Aeronautical Knowledge, FAA Manual 8083-25A, www.faa.gov

Spatial Awareness

It seems such a simple task to determine which way is up. Healthy humans are actually quite good at answering this question under ordinary circumstances. In fact, it's the critical function of our vestibular system that gives us spatial awareness and allows us to smoothly walk, run, jump, and otherwise maintain our balance. Our perception of up and down is based on sensing the influence of gravity's acceleration downward. We sense rotation through the relative motion of fluids in our inner ear as their inertia keeps them in place and our head rotates.

However, the aerospace environment is not an ordinary one. In flight, the aircraft routinely causes accelerations that make the occupants experience increased and decreased G-forces as well as perceived centrifugal forces. When pilots cannot maintain visual contact with the outside world, they are susceptible to spatial disorientation, which is an incorrect perception of the position and/or motion of the aircraft.

Spatial Orientation (Semicircular Canals, Disorientation)

On either side of our head are the semicircular canals (see Figure 8-23). The canals are a system of three fluid-filled tubes that are oriented at approximately 90° to each other. Each tube has a widening called the ampulla in which a gelatinous mass known as the cupula is affixed to a bed of small hair fibers that trigger the brain when they are pulled. When your body is stationary, there is no relative flow of fluid through the tube, so the cupula remains in place, and no signal is sent to the brain. However, if you twist your head, at least one tube roughly aligns with the direction of the rotation, and the walls of the canal rotate as the fluid within the tube remains in place due to its inertia. This relative flow pushes the cupula to one side, which triggers the brain that you are rotating in space.

For short duration motions, this works correctly; however, an airplane can experience a long duration motion in one direction, such as when in a slow turn.

This type of motion allows friction to cause the fluid in the canal to also start to rotate, at which point, there is no relative motion, and the brain is signaled that the rotation has stopped, leading to spatial disorientation. The situation becomes even more disorienting when the pilot checks the instrument panel, recognizes the motion, and quickly takes corrective action. The rapid correction results in relative fluid flow and the disturbing perception that the aircraft is actually turning in the opposite direction of the actual motion. For the pilot to change the aircraft's attitude enough to make the flight "feel right," the pilot has to return the aircraft to beyond the original orientation. This situation in instrument conditions can quickly become deadly if pilots do not know to ignore their biofeedback and perceptions and to trust their instruments.

Awareness of the position of our body is sensed through the pressure of gravity on parts of our bodies. We know we are lying down due to the pressure of the bed on our back, or that we are standing upright due to the pressure on the bottom of our feet. However, our primary sense for position is also sensed within the inner ear. The utricle and saccule both contain small chunks of calcium stones called otoliths (see Figure 8-24). The otoliths are attached to cilia hairs on nerve fibers, and when you tip your head to some position, the otoliths shift, triggering a signal to the brain. You can picture tipping your head up and backward, which results in otoliths being pulled backward inside the structure.

Spatial disorientation:

an inaccurate perception of the attitude and motion of the vehicle.

Figure 8-23 (a) The vestibular apparatus inside the human ear (b) components of the organs in the semicircular canals.

From STARR. *Biology: Concepts and Applications*, 7th edition. © 2008 Brooks/Cole, a part of Cengage Learning, Inc. Reproduced with permission. www.cengage.com./permissions

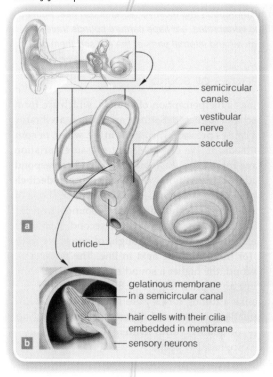

semicircular canals

vestibular nerve

saccule

a

utricle

gelatinous membrane in a semicircular canal

hair cells with their cilia embedded in membrane

sensory neurons

b

Figure 8-24 How otoliths move in response to gravity when the head tilts.

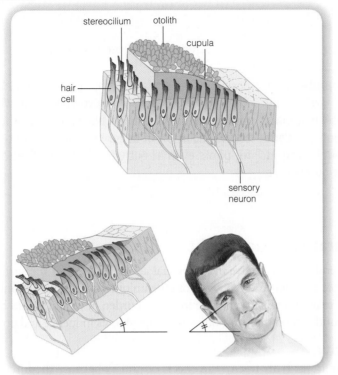

The danger is that a rapid acceleration forward such as is experienced during a takeoff, especially a catapult launch, would also result in the otoliths moving backward due to their inertia. Thus, a pilot being catapulted from a carrier deck experiences a **somatogravic illusion** and perceives that the aircraft is pitching violently upward. The corrective action would be to push the nose downward at the worst possible moment, just as the pilot is clearing the bow end of the flight deck. Instead the pilot needs to pull up as the aircraft clears the deck, clearly the required action is exactly opposite of what is called for based on the pilot's perceptions.

Fun Fact

Watch movies such as *Top Gun* that show aircraft being launched from a carrier deck, and notice what the pilot is doing just prior to the "cat" launch. How are the pilot's actions related to the illusion just described?

Due to the danger represented by spatial awareness illusions, pilots are trained and required to see and avoid instrument flight conditions unless they have been certified by earning their **Instrument Rating**, are flying an airplane properly equipped for flight in instrument conditions, and have logged sufficient instrument flight experience in the recent past to be current on the skills required to safely pilot the aircraft. The careful design, layout, and operational procedures for instruments can play a significant role in improving the pilot's spatial awareness and keep both human and aircraft safe.

GRASS STRIP ADVENTURE

Start up Flight Simulator. Take off and take a heading toward a distant airport. Now use the weather controls to create instrument flight conditions. Can you maintain control of the airplane for more than three minutes? Focus on maintaining level wings, altitude, and heading for a little less than the predicted flight time to your destination. You can turn off the weather or start a descent to landing to see how well you have done at flying in the soup. For even more fun, fly an instrument approach to your destination!

Long-Duration Space Flight

When we send humans to live in places that are different from the surface of Earth, they are liable to suffer in some way from the unique conditions. Although we might be tempted to assume that being free from the burden of weight would be liberating, we need to understand that our bodies are adapted to its influence and not its absence.

Microgravity

Humans are never truly free of the effects of gravity because the force exerts itself whenever two masses exist. Each pulls on the other with the same exact force, which we refer to as the weight of the object. However, we can experience **microgravity** by eliminating the result of gravity, which is the perception of this weight, simply be allowing all objects in the immediate surroundings to fall together. Because all objects experience the same acceleration due to gravity, regardless of their mass, when we don't try to maintain our altitude, there isn't any pressure between any two objects, and they all appear to be "weightless."

On Earth, we can very briefly experience microgravity by leaping off of a structure or flying in an aircraft that is piloted to follow the same parabolic path that an object in free fall would follow. NASA has used an aircraft lovingly referred to as the *Vomit Comet* for many years to support microgravity research without the high cost of space flight.

To achieve microgravity for longer durations requires placing a craft into the perpetual falling of an orbit around the planet. Some modern materials such as Aerogels, the best insulating material ever developed, would benefit greatly from manufacturing in reduced gravity environments. However, this setting is very stressful for the human body.

Gravity causes our body to experience significant impact forces between the bones of our skeleton every time we take a step. Although we might find this to be tiring after a long day of walking, it plays an essential role in stimulating the bones in our body to maintain their size and strength. When we don't expose ourselves to these impact forces, our bones shrink in size and become significantly weakened.

In addition, our muscles are also adapted to the world in which we live. We know that if we undertake weight training programs, we can increase the size and strength of our muscles, and that if we choose to lay around without physical exertion, the muscles will atrophy and loss mass and strength. Ask anyone that has had a cast in place for more than a few weeks about their experiences after the cast was removed. We very rapidly become weaker! In addition, when we don't stretch out our muscles, they tend to shorten in length. Watch NASA videos of astronauts in action, and you will note that they regularly float with their foot pointed rather than at its Earth orientation perpendicular to the leg. This effect is due to the reduced forces on the foot and leg and the resulting shortening of the calf muscles along the back of the lower leg.

None of these effects sound overly detrimental until it is time to return the person to the surface of Earth. If the body has lost its adaptations of muscle strength and length and bone density and strength due to long-term exposure to microgravity, it can be extremely uncomfortable or even impossible to safely return to normal Earth activities. Most astronauts require extensive physical rehabilitation upon return from space. This is one of the most significant challenges for long-duration space flight and colonization of other solar system environments and will require the development of ways to apply stress to our bodies to maintain our muscular and skeletal systems.

Psychology

Few humans have suffered the grueling work conditions of long-duration space flight. Spacecraft are small, cramped spaces, and missions are typically scheduled down to 5-minute blocks of time, including meals, rest, and mission activities. On Earth, comparable experiences include that of submariners aboard nuclear submarines (typical time submerged is 6 months) and participants in Biosphere 2 (Biosphere 1 is Earth itself) experiments in the Arizona desert outside of Tucson. Both represent closed systems in which the air, soil, water, and other materials are sealed within the environment for many months at a time. Travel to other planets will make these long durations seem insignificant as round-trip experiences to Mars may require more than 24 months sealed into a space with the same crewmates.

CHECKLISTS AND PROCEDURES

Every critical aspect of an aircraft's or spacecraft's operation is managed through the use of checklists and standardized procedures. The use of a checklist reduces the chances of missing an important observation, incorrectly using a piece of equipment, or maneuvering the craft in a way that is unpredictable to the rest of the aviation world or dangerous to the vehicle, its crew, or to their surroundings (see Figure 8-25).

Figure 8-25 *Pilots use checklists in all phases of flight to ensure the flights safety and reduce errors.*

Source: Pilot's Handbook of Aeronautical Knowledge, FAA Manual 8083-25A, www.faa.gov

Typical Checklists

Pilots routinely perform many preflight, flight and postflight checklists that are specific to the flight and aircraft. Often pilot error occurs when a checklist is ignored. Checklists are extremely important for both rookie and veteran pilots and cover all aspects of a vehicle's operation. Good pilots are routinely familiar with the vehicles they are operating.

Aircraft

The following are a few examples of flight checklists for aircraft. These can range from a few simple steps to several steps.

▶ Preflight Inspection

▶ Engine Start

▶ Taxi

▶ Preparation for Takeoff

▶ Takeoff

▶ Preparation for Landing

▶ Landing

▶ Engine Shutdown

Space Missions

Space mission checklists are just as critical as aircraft missions and can be very complex. This sample checklist from Apollo 12 is several pages and is outlined by minutes (see Figure 8-26).

Following are a few examples of space mission checklists:

▶ Vehicle Preparation

▶ Fueling

▶ Crew Insertion

▶ Ignition and Launch

▶ Extra Vehicular Activity

▶ Retro Burn

▶ Reentry

▶ Recovery

Figure 8-26 During space missions, the crew is kept on task by checklists such as this one worn by astronauts during the lunar landing of Apollo 12.

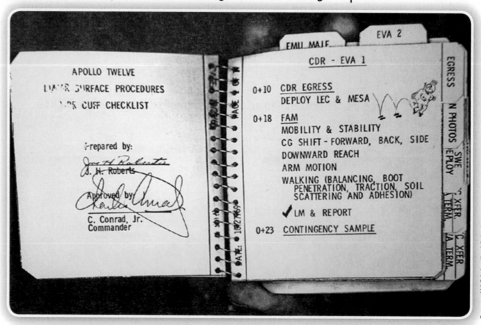

Source: NASA Apollo 12 Archive

Careers in Aerospace Engineering

A CAREER TAKES FLIGHT

"As a kid, I really liked science fairs, putting a project together and demonstrating it," says Kit Siu, who now helps put projects together at the General Electric Global Research Center in Niskayuna, New York. "And that excitement is still there," she says. "You're always asking yourself what the next cool thing is, then trying to convince your colleagues or your customers that it's worth looking into."

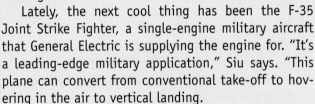

Kit Sui

© Cengage Learning 2012

Lately, the next cool thing has been the F-35 Joint Strike Fighter, a single-engine military aircraft that General Electric is supplying the engine for. "It's a leading-edge military application," Siu says. "This plane can convert from conventional take-off to hovering in the air to vertical landing.

"Making it work involves both interaction between the aircraft and the engine and a lot of advanced technology," Siu adds. "And this is just one of the areas that the research center is involved with. We're also looking at things that haven't been invented yet. In all these projects, it's really rewarding to see your design go from computer simulation all the way to the testing of real parts."

On the Job

A control system engineer, Siu keeps an eye on the big picture while zeroing in on the things that will make the engine perform the way it needs to. "It's kind of like designing a cruise control for your car," she says, "or a thermostat for your house. In each case, the system has to follow your command, and that means looking at the whole thing. For the thermostat, you need to know what kind of house it is, what kind of heating system it has, how many zones there are, the type of fuel. Once you've identified your system, you can model it. You take some part of the real world and create a simulation."

For the F-35, Siu and her team get input from the airframer and the military pilots who will be flying the planes. "They're concerned with the feel of the aircraft," Siu says, "how responsive it is to their commands. So, when they move the stick and throttle, we have to make sure we give them all the thrust they need to make their maneuvers while keeping the engine in safe operating condition—preventing stall and over-acceleration, for example."

Education

Siu has a BS in electrical engineering from Columbia University, whose engineering curriculum she remembers fondly for the diversity of its course offerings. "We were required to take a number of liberal arts courses, even though we were engineering majors," she says, "and I really liked the balance that was struck between science and the humanities."

Even as an undergrad, Siu had her sights set on working for a diverse company like GE. "I was accepted into their Edison Engineering Development Program, which takes undergrads fresh out of school and puts them through a series of rotations in various labs, all while they're getting their master's degrees." Siu completed the three-year program in 2003 with an MEng in computer engineering from the Rensselaer Polytechnic Institute.

Advice to Students

It may have killed the cat, but curiosity is what Siu recommends you hold on to if you're considering a career in engineering. "If you have a passion for figuring out how things work and how they might work better, you'll do well," she says. "Engineering is all about solving problems."

One problem that the engineering profession hasn't completely solved is the percentage of women who practice it. Siu says that even today she'll sometimes look around the room and she's the only woman there. Her response? "It's just something I had to get over," she says. "It tell myself, 'You belong here just as much as everybody else.'"

SUMMARY

- When we take our bodies up into the atmosphere or out into outer space, we are confronted by the limits of our physiology.

- We depend on the presence of oxygen in sufficient quantity and under enough pressure to force itself into our lungs for absorption and distribution throughout our entire body. When insufficient, we lose our ability to focus and think clearly, we gray out, black out, and eventually die if oxygen isn't available.

- Our hearts deliver the oxygen and energy that we need for every cell in our bodies. The heart is a pump that is limited in its ability to maintain blood flow against the increased accelerations created when an aircraft maneuvers.

- We are very sensitive to temperature and need to regulate our body core temperature within a very narrow range to remain functional. Vary our temperature just a little upward or downward, and we shut down.

- Human sight is extraordinary for its high resolution and color vision. However, the resolution is limited to a very narrow angle of the world in front of us, and our color vision fails as the intensity of illumination decreases.

- We lose all sense of up and down and motion whenever a motion is slow or uniform unless we maintain a visual tie to the outside world. When we lose sight of the world around us, we are easily disoriented.

- Travel in space presents the special challenge of microgravity. Our bones and muscles shrink, and our ability to carry out a normal life on Earth requires painful rehabilitation upon our return. No humans have ever been secluded as a group for as long as will be required to travel to even the closest of our neighboring planets.

- To overcome these limitations, we develop equipment that enhances our capabilities, procedures that safeguard us from foreseeable accidents, and checklists to make certain that no issue of safety is overlooked.

1. Create a timeline that explores the advancements in human physiological devices as necessitated by aircraft performance.
2. Create a small experiment you can do in class that addresses a particular physiological aspect.
3. Research the limits of human physiology and relate your information to piloting ergonomics. Compare and contrast different pilot postures and how they would benefit or hinder pilot performance.
4. Design an aircraft that would give a pilot maximum G-force turning.

5. Describe the types of training fighter pilots and astronauts endure to counter physiological effects.
6. Research the G-forces experienced by Space Shuttle astronauts as they journey from launch pad to orbit. Create a G-force graph for the time during the stages of the flight.
7. Explore and explain the conditions for which it is possible for water to freeze and boil at nearly the same time.

EXTRA MILE

1. Check your local TV guide or check online videos at www.redbullairrace.com for information on the Red Bull Air Race series to see pilots push the physiological boundaries. Take note of pilot behaviors to see how they compensate.
2. Create your own microgravity experiment related to human physiology.

3. Design a checklist for an everyday procedure such as driving or making a meal.
4. Design an experiment you would like to test on the zero-g parabolic flight trials.

CHAPTER 9

Aerospace Materials: Building for Strength, Weight, and Speed

| | START LOCATION | DISTANCE | END LOCATION |

Menu

Before You Begin

Think about these questions as you study the concepts in this chapter:

1 What materials are used in the manufacture of an aerospace vehicle?

2 Why was wood still common after metal aircraft had been designed (see Figure 9-1)?

3 Why did aluminum rapidly replace wood for most aviation applications (see Figure 9-2)?

4 What is material "fatigue," and why is this critically important in engineering?

5 How is an airframe constructed out of aluminum?

6 What is a composite material?

7 When composite materials age, how are weaknesses identified and repaired?

8 How are ceramic materials used in the construction of aerospace vehicles?

9 What emerging materials are changing how aerospace vehicles are designed and built?

10 How are materials tested and evaluated?

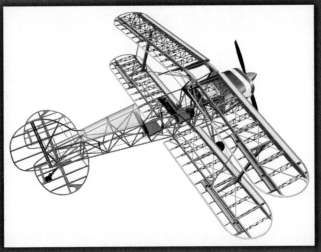

Figure 9-1 *Airframe for a Skybolt aircraft without fabric covering. The Skybolt has a welded metal airframe and wooden wing ribs and spars.*

Image by Steen Aero Lab

Figure 9-2 *Complete kit for the home-built Sonex with an all-aluminum airframe.*

Courtesy of Sonex Aircraft, LLC

Throughout this chapter, our emphasis will be on the broad impact that materials science has on the design, construction, maintenance, and safety of an aerospace vehicle. Humans are tool users. Throughout our collective experience as a species, we have sought to understand and control our environment to improve our standard of living. We have harvested and processed naturally occurring materials to construct weapons, protective clothing, shelter, storage systems, medicines, communication mechanisms, and transportation devices (Figure 9-3). In all cases, we have sought to take advantage of the material's inherent strengths and beneficial uses while minimizing the impact of their weaknesses and damaging potential. In the past century, we mastered the ability to manipulate the structure of the natural world to create entirely new materials that offer even greater benefits or more limited restrictions on use (Figure 9-4).

Figure 9-3 *Workers assemble Bristol Bulldog Fighters in 1930.*

Figure 9-4 *Brazilian workers assemble the complex riveted aluminum airframe for a commercial airliner.*

For the modern aerospace engineer, this has meant that we can choose to use wood, metal, plastic, composite, and ceramic materials in ways that were impossible even a few decades ago. So why have metal and composite materials almost entirely replaced wood in the airframes and skin of our vehicles? What are the advantages and disadvantages of each of these materials? To understand the decisions that are made during the design of an aerospace vehicle, we need to understand the answers to these questions.

TYPES OF MATERIALS

A wide range of materials has been used for the design and manufacturing of aerospace vehicles, each with its own unique advantages and disadvantages. In this section, we explore the common materials that have played a significant role in aerospace as parts of the airframe of the vehicle.

Wood

Wood was the first readily available aerospace building material. Wood is a naturally renewable resource that is readily available almost everywhere that humans maintain permanent communities. Wood from ash and hickory trees were some of the first woods used in the manufacture of aircraft; however, they were quickly replaced by Sitka spruce. Spruce is the wood of choice due to its relative abundance, widespread availability, high strength, and low weight in comparison to other woods. Sitka spruce in particular is relatively consistent for a natural substance, meaning you can make parts such as ribs or spars from several sources of trees, and they will be nearly the same weight and strength. Although wood is not as strong as steel or aluminum, by shaping the wood and building it into simple structures with time-tested construction methods and structural techniques, wood is comparatively strong for the amount of weight that can be saved. Our long experience with working with wood also means that there is a large pool of skilled labor to draw upon for the construction of aircraft. Anyone that can build cabinets or furniture is competent to construct an airplane out of wood (Figure 9-5).

Wood has another very desirable characteristic in that it responds well to being flexed repeatedly and to vibration. During flight, the forces required to fly and maneuver the vehicle are frequently changing. The components of the aircraft flex farther from their neutral position when the forces increase and relax back to their neutral position as the forces are removed. This continuous flexing is absorbed by the fibers of the wood as it stretches and relaxes but remains intact. This makes wood an excellent material for structures that experience a lot of long-term exposure to cycles of flexure (Figure 9-6). Because an airplane is suspended in the air, it isn't braced by the ground. This means that the vibration from engines, propellers, and air flowing over the skin of the craft all continuously shake the vehicle. Wood acts as a natural shock absorber to **dampen** out and reduce the amount of vibration felt. With the change from reciprocating piston engines to jet engines, there has been a drastic reduction in the amount of vibration produced by the aircraft's engines.

Figure 9-5 Fuselage, cockpit, and powerplant detail of a wooden *Albatross* Fighter of World War I. Drawing by Mark Miller.

Mark Miller

Courtesy of The Aerodrome

Figure 9-6 Wing structure of the wooden *Albatross* Fighter of World War I. Drawing by Mark Miller.

Courtesy of The Aerodrome

Figure 9-7 A restored World War I wooden aircraft from the Sopwith Company.

Figure 9-7 A restored World War I wooden aircraft from the Sopwith Company.

Experimental category:

an airworthiness certification for an aircraft that has been constructed by amateur builders who have documented that they performed at least 51% of the assembly work to manufacture the vehicle.

Reinforcement:

a material incorporated into a structure to add strength.

Wood has remained a popular choice for builders of **experimental category** (homebuilt) aircraft because the skills and techniques required to work with wood are easily learned (Figure 9-7). Wood construction is found in spars, ribs, floorboards, instrument panels, and as **reinforcement** for carbon fiber or fiberglass composites.

Most aircraft with wooden frames use a cloth covering for the fuselage and control surfaces. This means the bulk of the strength of the fuselage relies on the strength of the wood frame much like the roll cage on a racecar. It also means that the fabric covering has to be firmly attached to the airframe so that aerodynamic forces don't make it vibrate and separate from the airframe (Figure 9-8). Common fabric attachment techniques include stretching, stitching, shrinking, gluing, and nailing the fabric in place (Figure 9-9). Once covered, an aircraft with a fabric skin requires multiple coats of **UV (ultraviolet)** blocking material and final and trim coats of paint (Figure 9-10).

Figure 9-8 Fabric can be glued to a metal frame with a special adhesive.

Figure 9-9 Stitches pass through the upper and lower cloth to secure it to the wing ribs.

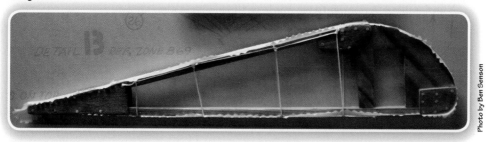

Photo by Ben Senson

Figure 9-10 Finishing fabric involves multiple steps: A) bare fabric, B) aluminized UV protective coating, C) color coats applied until the proper color depth is produced, D) trim colors applied.

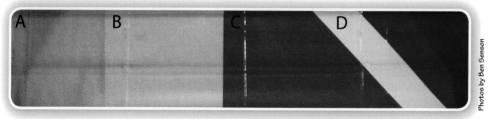

Photos by Ben Senson

Figure 9-11 The *Spruce Goose* with its all-wooden fuselage is the largest aircraft to have ever flown.

Source: Keystone Features/Stringer/Hulton Archive/Getty Images

The largest plane ever built was almost entirely of spruce wood (Figure 9-11). Contracted by Howard Hughes and tested on November 2, 1947, the *Spruce Goose* flew for a little over a mile at just 70 feet above the water during a taxi exercise. The *Goose* never had another flight and is now on display at the Evergreen Aviation and Space Museum in McMinnville, Oregon.

Building components such as the two sides of the fuselage or the multiple ribs for the wing require some special techniques for working with wood. Builders commonly first lay out a jig, which acts as the pattern for many of the same components with nearly identical dimensions (Figure 9-12). The jig allows components to be accurately dimensioned, placed, and bonded to each other. Once assembled,

Jig:

a mechanical device that aligns individual parts for assembly into a larger component.

Figure 9-12 The layout jig is made with a pattern for the component and small wooden blocks that hold all the pieces in place while they are being dimensioned and assembled. A complete rib made with this jig is also shown.

Photo by Ben Senson

Gusset plate:

a reinforcing brace attached to one or more sides of a joint to strengthen the joint and brace its alignment.

the components are strengthened with small **gusset plates** of thin plywood that reinforce both the bond strength and the alignment of all of the major wood joints in the component (Figure 9-13). Additional small blocks of wood are used to finish the shape for leading and trailing edges or attachment points for other aircraft components (Figure 9-14).

Figure 9-13 Gusset plates reinforce all the major joints in a wooden component such as this tail section of a fuselage.

Photo by Ben Senson

Figure 9-14 The leading edge of this horizontal stabilizer has been shaped with small wooden blocks.

Photo by Ben Senson

To summarize, the advantages of wood include the following:

▶ Easy to work and repair

▶ Lightweight

▶ Can be bent or molded

▶ Complex structures can be made to increase strength

▶ Will not fatigue (see the next "Metal" section for a description of "fatigue")

▶ Dampens vibration

▶ Renewable resource

The disadvantages of wood include the following:

▶ Inconsistencies from piece to piece

▶ Increasingly difficult to find aircraft-quality woods

▶ Increased costs for cutting too small

▶ Hard to recycle

▶ Vulnerable to insects and cycles of temperature and humidity

Metal

Metals are just about equally strong no matter which way you bend or push on them. This property makes them a great building material. One of the most popular metals used in **alloys** for aerospace is aluminum. When mixed with the proper amount of copper, manganese, and magnesium, aluminum becomes duralumin. Duralumin enhances the natural lightness of aluminum by greatly increasing its resistance to tearing. This makes it a great material for use as the skin of the aircraft, especially for aircraft where the skin provides a significant part of the airframe's strength as in most modern aircraft designs (Figure 9-15). The disadvantage of this alloy has been its susceptibility to **corrosion**; however, a thin coating of pure aluminum on the outside of a component to produce Alclad greatly reduces this tendency. Alclad is lightweight, durable, corrosion-resistant, and recyclable. Another corrosion-proofing technique is to prepare and paint the aluminum (Figure 9-16).

Figure 9-15 **The aluminum skin of a Cessna aircraft provides much of its structural strength.**

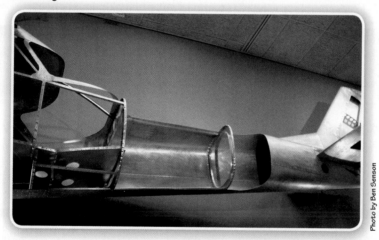

Photo by Ben Senson

Figure 9-16 **Finishing metal involves multiple steps: A) bare aluminum, B) acid-etched surface, C) corrosion-proof coating, D) color and trim colors applied.**

Photos by Ben Senson

Initially, metal tubing was introduced as a replacement for wooden internal structures, but engineers who were using sheet metal as a covering rather than cloth found they could greatly reduce the amount of support structure needed for the fuselage. This reduced the overall weight and made the aircraft more durable to weather, flight stresses, or airborne obstructions. As we learned earlier, reducing weight leads to an increase in flight performance, range, and maneuverability. Thus, many aircraft have an aluminum skin covering an aluminum airframe. When strength and temperature resistance are more important than weight reduction, other metals such as steel or titanium are used to make components.

Metal **fatigue** is one of many modes of failure for a material that experiences cyclical adding and removing of forces. When sheets of metal became common components in aircraft designs, the engineers understood that the material could corrode, buckle, wear out, tear, yield, or even rupture or fracture. However, when a material is repeatedly flexed due to vibration, pressurization, or varying aerodynamic forces, the material develops imperfections that greatly reduce the maximum force the component can tolerate before failing. Metal fatigue is progressive and requires regular inspections to identify its start and requires maintenance or repair of fatigued parts to ensure safety.

Building a structure out of aluminum can involved individual structures made out of a single piece of aluminum cut to size and stamped into shape (Figure 9-17) or from individual pieces **welded** into a frame (Figure 9-18). Structures made out

Figure 9-17 **A complete wing rib stamped out of an aluminum sheet.**

Figure 9-18 **Welding of individual components together makes for a very strong frame.**

of multiple identical components such as wing ribs are formed by creating a pattern shape from which to trace the piece and then bending the piece over a form block. The shape is then drilled and assembled (Figure 9-19). A full wing panel can be assembled out of only a few different components **riveted** together (Figure 9-20). Many components require multiple layers of material so that the strength of the overall component can be varied while reducing weight to a minimum (Figure 9-21).

Figure 9-19 Nearly identical wing ribs can be created: A) by using a pattern to cut the sheet aluminum, B) bending the blank over a forming block, C) drilling holes, D) securing the wing skin and other components with rivets.

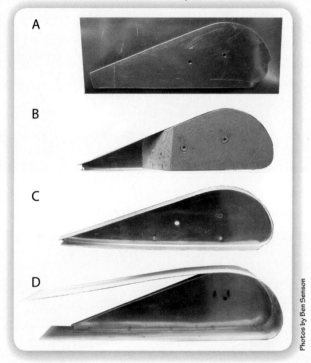

Figure 9-20 The wing panel from a Sonex is formed from a few components bonded to each other by pop rivets and driven rivets.

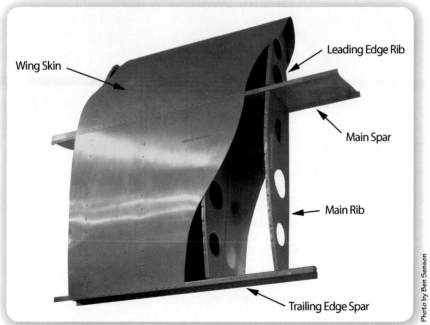

Figure 9-21 *The main wing spar from a Sonex incorporates multiple layers of aluminum to maximize strength where it is needed while minimizing the weight of the component.*

Photo by Ben Senson

Figure 9-22 *An example of an all aluminum aircraft is the Mahoney Sorceress.*

Photo by Ben Senson

The advantages of aluminum include the following (see Figure 9-22):

▶ High strength-to-weight ratio

▶ Low cost and widespread availability

▶ Proven durability and resistance to Sun and moisture

▶ Easy to work with, easy to repair

▶ Metal kit aircraft construction simplified by modern blind rivet fasteners

▶ Easy to form into many shapes, with almost no limit to the shapes it can be formed into

- ▶ Environmentally friendly
- ▶ Lightning protection provided by the outer shell
- ▶ Energy absorption in low-speed impacts rather than splintering or shattering
- ▶ Recyclable

The disadvantages of aluminum include the following:

- ▶ Subject to fatigue
- ▶ Hard to detect micro fissures and cracks
- ▶ Can buckle when compressed
- ▶ Corrosion on hard-to-reach areas

Composites

Aluminum is still found in abundance on current aircraft. However, advances in fiberglass and carbon fiber **composites** have pushed composite materials into the mainstream. A composite material is a compound material that is composed of two or more components that have significantly different properties and that remain distinct and separate from each other in the final structure that is made from the material. The most common composites in aviation have **fibers** that provide strength embedded in a matrix of **resin** that holds the fibers in place (Figure 9-23). Most new aircraft designs incorporate carbon fiber or even Kevlar composite assemblies in the fuselage, wings, or empennage. Some are now making entire fuselages from carbon fiber composites. The main reason carbon is such a popular choice is its incredible strength-to-weight ratio.

Composites are not just limited to carbon fiber or Kevlar. Plywood, fiberglass, and other similar products are considered composite materials. A composite is usually two components: a matrix and reinforcement. The matrix could be loose-fill fibers or woven fibers where reinforcement is a bonding material such as epoxy. The combined strength of the two materials exceeds the sum of the parts individually.

Composite:
a compound material that is formed out of two or more materials that remains distinct and separate components of a final structural piece.

Figure 9-23 *Fiberglass forms a composite material when embedded in a resin. The strength of the material comes from the number and orientation of the fibers while they are held in place by the resin.*

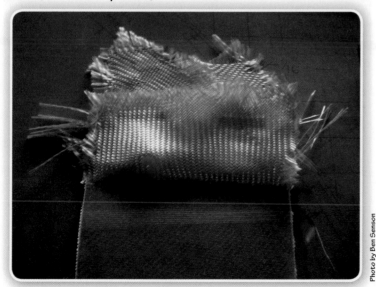

Photo by Ben Senson

Unlike metals though, how you apply the force and stress to composite materials does impact the strength of that part. Fiber orientation is very important in composite materials (Figure 9-24).

Engineers control the direction of strength in a component made with composites by varying the type, number, and direction of fibers within the component and by selecting an appropriate resin in which to embed them. Building with composites involves relatively few materials and is an easy skill to master (Figure 9-25).

Figure 9-24 Fiberglass comes in many weaves that vary in the number of fibers in any given direction. Shown here are BiD (bidirectional) in which nearly identical numbers of fibers are woven at 90° to each other and Uni (unidirectional) in which almost all of the fibers run in the same direction with just enough crossing the material to hold the fibers in place.

Figure 9-25 Home builders constructing the VariEze, one of the first all-composite aircraft, used unidirectional and bidirectional fiberglass and resin with and without a foam core.

Resins include polyester-based resins that require protection from UV radiation to retain their strength and need to be sealed to prevent degradation. Vinylester resins are more resistant to degradation and are more flexible. Epoxy resins are very strong and nearly transparent when fully cured, so they are used as structural glue between components. Shape memory polymer resins include many types of resins that have the unique characteristic that they can be reheated above a glass transition temperature (different for each resin), reshaped, and then cooled without changing their material's properties.

Composite materials have very high strength-to-weight ratios and offer the ability to mold a component into almost any shape. They can also be molded around incredibly lightweight core materials such as closed-cell foams to create seamless, strong, lightweight, high-performance aircraft components such as wing panels (Figure 9-26 and 9-27). These structures can include additional components such as wing spars, fuel tanks, landing gear wells, and so on.

Composite materials can fail due to the breakdown of the resin matrix, delaminating of one layer of fiber from another, or fiber pullout of single strands from the matrix. One issue with composites is that they can fail microscopically without visible failure. This requires maintenance procedures that include ultrasonic, thermographic, x-ray, and other nondestructive tests.

Figure 9-26 The canard wing core for the VariEze and one of the end templates used to cut the foam with a hotwire from a block of foam.

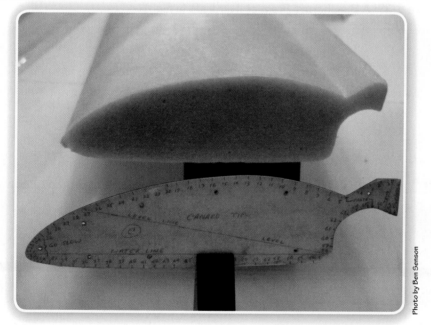

Figure 9-27 A cross-sectional cut through a composite wing showing the use of two different foams and a composite spar to strengthen the wing.

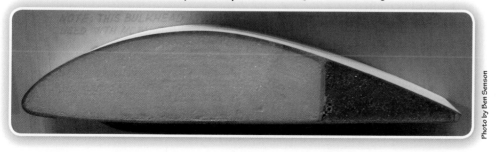

Figure 9-28 A Beechcraft Premier aircraft fuselage makes use of a more expensive carbon fiber weave for greater strength.

Courtesy of Hawker Beechcraft Corporation

Figure 9-29 A Beechcraft Premier II aircraft in flight.

Courtesy of Hawker Beechcraft Corporation

The advantages to composites include the following (see Figures 9-28 and 9-29):

▶ Easily repaired

▶ Superior weight to strength: 1/5 the weight of steel for the same strength

▶ Can be molded to many different shapes

▶ Can apply layers specific to stress needs (see Figure 9-30)

▶ Noncorrosive

▶ High vibration/sound dampening properties

Figure 9-30 Fiberglass is used extensively throughout the award-winning custom *Nemesis* aircraft.

Photo by Ben Senson

The disadvantages to composites include the following:

▶ Delaminating; layers start to separate

▶ Must be manufactured in clean environment

▶ Requires routine inspections using specialized equipment to detect microscopic failures

Ceramics

A ceramic material is created by heating and cooling an inorganic, nonmetallic material. The internal structure of a ceramic may include the formation of crystals or may be uniform such as in a glass material. Ceramics tend to be strong, but they are not "tough" because they don't bend much before breaking. Most ceramic use is limited to compression applications such as bearings or on certain types of vehicles as heat shields, such as on the Space Shuttle (Figure 9-31).

Ceramic:
a material that is created by heating and cooling an inorganic, nonmetallic material. The material can contain crystals of the material or be glassy in composition.

Figure 9-31 *Close up of Space Shuttle Endeavour showing the ceramic tiles that provide its thermal protection during reentry.*

Courtesy of NASA

Ceramics are not only strong, but they also can be electrical and thermal insulators. The heat tiles on the Space Shuttle are a little more than 2½ inches thick and 6 to 8 inches square but weigh very little (less than 0.75 lbs) for their size. These special types of ceramics have a lot of air mixed in with them. This trapped air acts as a great insulator that won't burn away like fiberglass at high temperatures.

Recycling Materials

Throughout the lifetime of an aircraft, various components will degrade or be damaged and need to be repaired or replaced. At the end of the aircraft's life, the entire airframe is declared nonairworthy and is generally scrapped. Aerospace engineers need to consider the disposal of all of the materials used in the manufacturing of the aircraft during their useful life and at their end of life. Disposal of materials is a major issue especially as the materials evolve from easily recyclable wood and aluminum to more durable materials such as rubber, composites, and ceramics. These last materials can be used as playground groundcover, road base, or concrete fillers. For metals in particular, it is much more energy efficient to recycle existing metal components rather than mining, purifying, and manufacturing new metals.

FUTURE OF MATERIALS

Many materials are currently considered exotic because they are not yet available for use in commercially manufactured vehicles. However, these materials show tremendous potential for helping aerospace engineers solve vehicle design problems.

Smart Metals

Smart metal:

after this material is formed, it can alter its specific shape while remaining a solid based on changes in temperature, electrical currents, or other energy inputs.

A smart metal material is capable of causing motion when activated. The materials have fairly limited range of motion but when combined into larger structures, they are capable of reshaping, or morphing, so the entire shape of a wing panel is transformed to maneuver the aircraft or for operations at low versus high speed (Figure 9-32). On a smaller scale, these smart metal materials are capable of self-healing, which may ultimately result in fatigue proofing aerospace structures.

Carbon Nanostructures

Carbon nanotubes are no less amazing for their incredible strength and electrical conductivity properties (Figure 9-33). Use of carbon nanotube structures in aerospace materials can enhance the lightning strike protection, reduce susceptibility to structural fatigue, and dampen vibrations. Another interesting property of carbon nanotubes is their ability to act as strain sensors within the active component. This may allow active monitoring of fatigue. Another application involves reshaping the material to create crash energy-absorbing structures. If we become adept at weaving carbon nanotubes into long tethers and cables, they are capable of providing the required strength-to-weight ratio to construct space elevator prototypes. Self-healing skins may also be possible with carbon nanotube materials.

Figure 9-32 NASA's 21st Century Aerospace Vehicle concept makes use of morphing materials and characteristics to change its shape in flight.

Courtesy of NASA Dryden Space Flight Center

Figure 9-33 Carbon nanotubes can be manufactured so that they can fit inside of other nanotubes. In this diagram, nanotubes vary in diameter from 4 nm to 20 nm in diameter. Notice that all of the tubes are formed from the hexagonal bonding pattern of the carbon atom.

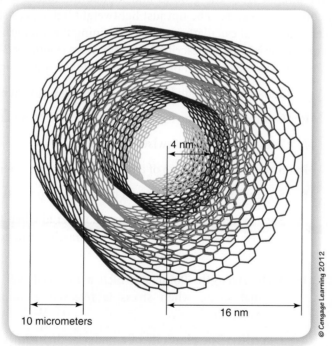

4 nm

10 micrometers

16 nm

© Cengage Learning 2012

Carbon nanofibers woven in composite materials produce much higher fracture toughness and reinforce the material for a higher delaminating resistance. Carbon nanofibers can also be assembled to create improved filtration and thermal insulation characteristics. The electromagnetic properties of carbon nanofibers allow them to be used for interference shielding and electrostatic discharge shielding. In aerospace structures, carbon nanofibers allow designs with extremely high stiffness and high thermal conductivity but at greatly reduced weight (see Figure 9-34).

Figure 9-34 A NASA scientist uses a chemical vapor deposition chamber to manufacture carbon nanotubes. The glowing light is the gaseous plasma of ionized carbon atoms that are used to form the carbon nanotube.

Courtesy of NASA

TESTING OF AEROSPACE MATERIALS

One of the most critical considerations when designing an aircraft is weight. However, to lift the aircraft from the ground and handle the forces of maneuvering flight, every component needs to be not just lightweight but also as strong as possible. Engineers use several calculations for determining the strength of materials. The strength of a material can be described in many ways (Figure 9-35).

Common material properties include the following:

Figure 9-35 Wing load testing of a Xenos motor glider. Note that the wing is mounted inverted so that the sand bag loading tests the application of positive G-load on the actual aircraft in flight. In the upper frame, the wing panel is loaded with 690 lbs (40% of ultimate load) while the lower frame has 3,710 lbs (103% of ultimate load).

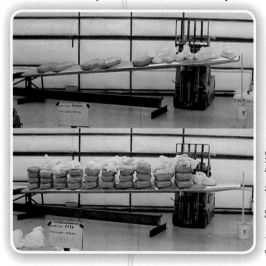

Courtesy of Sonex Aircraft, LLC

▶ **Tension:** Associated with forces pulling outward on a material.

▶ **Compression:** The forces pushing inward on a material.

▶ **Torsion:** The forces twisting a material.

▶ **Stress:** The resistance of a material during an external load.

▶ **Strain:** The deformation of a material when stress is applied.

▶ **Shear strength:** The strength of a material when the forces are applied parallel to each other and mostly perpendicular to the material.

▶ **Yield strength:** Point at which a material will return to its original shape when stress is released without permanent deformation.

▶ **Ductility:** The plasticity of a material that allows it to bend or form before fracture.

▶ **Fatigue limit:** The point of failure caused by cycles of stress applied to a material.

▶ **Malleability:** The ability of a material to be formed or bent under pressure without fracturing.

▶ **Young's modulus:** The relationship of stress over strain.

Aeronautics

Recall from earlier chapters that an aircraft must have a medium, such as air, to operate. The large control surfaces and propeller of a single-engine aircraft would be useless in the vacuum of space. As aircraft fly and maneuver through the atmosphere, many types of stresses are placed on the airframe and control surfaces. As explained earlier, materials need to be strong but lightweight. Many of the determining factors for materials used in a particular aircraft depend on the purpose of the aircraft. A traditional general aviation aircraft is usually constructed out of common materials such as aluminum and fiberglass. Lightweight, high-speed aircraft usually require more modern materials such as carbon fiber and titanium. This is not always the rule, but the point is to understand the role of the aircraft first and then to select cost-effective materials based on that. You could make a general aviation aircraft from carbon fiber, for example, but the cost would be prohibitive.

Now that we have a better understanding of "strength," you can see how some materials might work better for wings than others. If you've ever sat in a window seat near the wing of a commercial airliner, you might have noticed some flexibility in the wings as the plane encounters turbulence. The wing tips of some airliners can flex more than 10 feet before failing. How is this possible? Engineers use various metals and structural designs to allow for some flexibility in the wing. Too much flexibility, and the wings might drag on the ground (Figure 9-36). Not enough flexibility, and the wings could shear off in turbulence.

Another consideration is for the pilots and passengers. The first pressurized cabin wasn't introduced until 1940 on the Boeing Stratoliner. This introduced a new component to aircraft stresses that became very evident in later aircraft. As an aircraft cabin is pressurized, it expands slightly. When it returns to "normal" atmosphere, the cabin shrinks back to its original size. This is not something you can easily observe because the expansion and contraction is a matter of millimeters, but this does cause fatigue on the aircraft.

Control surfaces are another consideration for aircraft. In the early years of aviation up until the early 1940s, cloth was the material of choice for covering wings and the empennage. It was lightweight and provided sufficient

Figure 9-36 A Boeing 787 wing flex test bends the wingtips upward nearly 25 feet (7.6 m) above their neutral position under a load that is 150% of the maximum load the aircraft should ever experience in flight.

Courtesy of Boeing Corporation

Fun Fact

Lessons Learned from the Comet

One of the most notorious incidences of metal fatigue and failure was with early models of the DeHavilland *Comet 1* (1950s). After several early accidents and two major in-flight breakups, engineers conducted pressurization tests on the *Comet 1*. They found that the stress buildup in the metal around the corners of the square windows was much higher than expected and could lead to explosive depressurization when the component failed due to metal fatigue, resulting in the breakup of the aircraft. The windows were redesigned with oval windows for the *Comet 2*—a design that is seen today in many modern commercial aircraft.

manipulation of the airflow. In later years, as speeds increased and more demands were placed on the airframe, thin sheet metal such as aluminum was used. This added rigidity and strength to the lift and control surfaces but at the cost of weight.

Propeller designs changed as a result of improved materials. Rather than use laminated carved wood that was heavy and expensive, propellers were made from hollow reinforced aluminum. This change reduced the weight and decreased the torque effect (wasted energy) the engine had to overcome.

Astronautics

As speeds and forces on structures increased, material needs for aircraft also increased. Much of what was learned from the supersonic test flights of the 1950s and 1960s helped pave the way for spacecraft materials. Engineers learned how to produce lightweight materials that could handle the high heat caused from friction with the atmosphere. New metal alloys that use titanium were created that had incredible strength-to-weight ratios, and existing aluminum alloys were improved.

Designing a craft for space is much different from a traditional aircraft. First, there is no atmosphere in space, so there are no control surfaces. Second, the craft is traveling at extremely high speeds to maintain orbit, but there is no atmosphere to cause friction until the craft tries to reenter the atmosphere and must have a protection system from the heat. Finally, there are micro-meteorites and higher levels of radiation that must be dealt with. For those of us on the ground, Earth's atmosphere protects us. But in space, even an object as small as a grain of sand can cause serious damage to a craft traveling more than 17,000 mph (a typical LEO speed)!

On the sunny side of an object, solar radiation can raise temperatures in excess of 235 °F while the shaded side will be below −100 °F.

Fun Fact

Aluminum Foil, Mylar, and the Moon

When designing the Apollo missions Lunar Excursion Module (LEM), the engineers were faced with the need to add thermal protection to the spacecraft without adding significant weight. Their solution was to use a relatively new, highly reflective material, called Mylar. Wrapping the entire spacecraft in multiple layers of the material only added pounds to the craft but provided the necessary protection. The LEM looked like a TV dinner (Figure 9-37)!

Figure 9-37 The *Eagle* lunar module shows its Mylar thermal protection system on the descent stage. In the lower corner is a close-up of a similar product that blocks 99.999% of radiant energy.

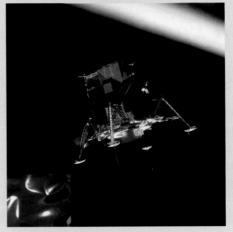

Courtesy of NASA

Careers in Aerospace Engineering

AN ENGINE FOR CHANGE

"As a very young child, I always took apart anything I could get my hands on," says Chukwueloka Umeh, who figured out how to reassemble everything he took apart as he got older. "In my parent's home, I went from being called 'Destroyer' to 'Mr. Fix It.'"

Umeh's childhood curiosity about how things work and how to make them function more effectively has continued to guide the native Nigerian in his work today. Currently employed by GE Aviation, Umeh has spent five years guiding part of a team responsible for designing engines that will propel aircrafts farther and burn less fuel while polluting the environment as little as possible.

Umeh's current job title is Engineering Manager for Commercial and Military Engines/CDN Module. The job entails overseeing a 27-person team responsible for designing the Combustor, Compressor Diffuser, and High-Pressure Turbine Nozzle (CDN) of aircraft turbofan engines. It is in this module that air and fuel are mixed, burned, and delivered to the high-pressure turbine. The process of burning the fuel and air generates a large increase in energy that accelerates the air flowing through the engine, thus producing the thrust that pushes the airplane forward.

Previously, Umeh was a scientist at GE Global Research Center in Niskayuna, New York, where he developed technologies for gas turbine engines to help reduce their harmful exhaust emissions as much as possible. He also worked on the Pulse Detonation Engine program, a novel engine with the potential of propelling an aircraft at over three times the speed of sound.

"At the research center, I was working on technology and science," Umeh says. "Now, I'm using the results of years of research and development to design and build greener, more efficient jet engines. It is very rewarding to see and touch the finished product you helped design. I love every moment of it!"

Eloka Umeh

© Cengage Learning 2012

On the Job

Whereas his research days involved running lots of experiments in a laboratory, accumulating and analyzing data (and often providing recommendations to design teams based on that data), Umeh's current work involves reviewing new engine designs, reviewing ideas to improve existing designs, and fixing problems encountered by airlines.

"What I like about being an engineering manager is that every day is different, and I get to work with engineering teams in different countries," he says. "When I first get to work on some days, I go straight into a meeting. On other days, I may spend more of my time brainstorming new design ideas. Most days, my time is shared between working on designs, reviewing designs with different design teams, dealing with schedules and budget, and meeting with different teams, including customers. There is never a dull moment at work, and I'm kept very busy from when I get to work to when I leave the office.

"In my current job, the customers that my team works with consist of the different airlines that use GE's engines on their airplanes. We also work with airplane manufacturers like Boeing, Airbus, Embraer, and Bombardier, to name a few."

Inspiration

As an engineer accustomed to dreaming up new ideas for different products, Umeh believes that to achieve anything great, one must both dream big and work hard.

Education

Umeh has a BEng in Electronic Engineering from the University of Nigeria, a BS in Aerospace Engineering from Embry-Riddle Aeronautical University in Daytona Beach, Florida, and an MS and PhD in Aeronautical Engineering from the Rensselaer Polytechnic Institute in Troy, New York.

Advice

Don't worry too much about which engineering discipline to study, Umeh says. "The important thing is to identify where your passion lies and follow it. Engineers are trained to solve problems following basic scientific procedures, using the tools available to them. This makes them valuable in any field they choose to go into, including aviation, medicine, transportation, petroleum, banking, science and research, among others. Great engineers and inventors are never afraid to dream big. No problem is too great to solve. It just takes time, resources, and teamwork."

SUMMARY

- Aerospace vehicles have been constructed out of a number of different materials. Common aerospace materials include the following:
  - Wood
  - Metal
  - Composites
  - Ceramics
- Advanced materials include the following:
  - Smart Metals
  - Carbon Nanofibers
- The goal of the aerospace engineer is to understand the strengths and weaknesses of all of the materials from which a vehicle can be designed and then to choose the best material available for the task required. The material selected for a component is constrained by a large number of characteristics:
  - Strength in tension, shear, and compression
  - Density
  - Flexibility
  - Ease of shaping the material
  - Flammability
  - Temperatures of phase changes
  - Resistance to fatigue
  - Failure modes
- Designs are developed, prototyped, and tested under simulated and real-world conditions to ensure that the product meets or exceeds the design requirements. This final stage of testing is critical for ensuring the safety of the final design.
- As material science continues to advance, the aerospace engineer has an ever-increasing variety of materials to choose from in creating aerospace vehicles of the future.

BRING IT HOME

1. Calculate the number of times that the skin of a jet is flexed and relaxed if it flies 20 trips a day between the islands of Hawaii over a 20-year time period.
2. Research the average number of flights a typical commuter jet flies in a day and how long they remain in service.
3. If an aircraft is designed for a "typical" service life, how soon would we expect it to fail in operations with the Hawaiian airline?

EXTRA MILE

1. Visit an aviation mechanic, and set up a tour of their aircraft repair center. Note all of the different materials that are used in the construction of the aircraft.
2. Research the term "corrosion," and consider its effect on an aircraft. How is corrosion prevented?
3. Construct various small wing spars out of wood, metal, foam, or composites, and design a method for testing them for strength. Consider using an equal weight of material versus using identical shapes of materials, and note the effect on the strength of the material.
4. Construct a model aircraft out of wood. Notice the strength of the materials as raw materials after they are assembled into a finished structure, such as a wing panel, and after they are covered by a skin.

CHAPTER 10
Remote Systems

GPS DELUXE

Menu

| START LOCATION | DISTANCE | END LOCATION |

Before You Begin

Think about these questions as you consider the concepts in this chapter:

1 What are the pros and cons of using remote systems versus direct human observation or control?

2 What are the major benefits and limitations of the various types of remote systems?

3 How have communications technologies improved remote systems?

4 What are some programming techniques commonly used in the field of aerospace engineering?

5 Why is it important to select the right hardware and software for a robot or system of robots to perform a specific set of tasks?

6 How can sensors be used to control a system?

7 What role do output devices play in remote systems or vehicles?

8 What is the significance of open-loop or closed-loop control?

9 What is the importance of systems engineering, and how might it affect design?

10 What benefits do robotics competitions offer to the industry or individuals?

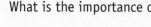

We have grown so accustomed to remote devices in our daily lives that it hardly fazes us when the latest micro RC plane or car is advertised on our remote operated television. You can control your DVR at home through your cell phone or the Internet. Doctors are performing surgeries from thousands of miles away on tele-robotic operating tables. Military and civilian police units use remotely operated devices in hazardous situations on a daily basis. In aerospace, we often use remote systems to investigate, explore, and engage unknown areas. Remote systems have been in use for decades.

In essence, aren't humans' just complex robots (see Figure 10-1)? We have sensors like robots (or rather we have given robots sensors like our senses)—eyes to see, ears to hear, nerves to feel. But robots can be equipped with sensors that exceed the capabilities of our human senses; they can communicate almost instantaneously with other robots, can detect light waves or radio waves beyond the spectrum we can see, can very accurately determine temperature, and can perform billions of calculations a second. Despite all the advanced sensors, remote systems still lack the ability to make freewill decisions. Humans do not always make good decisions, but when it counts, we can evaluate a situation and step out of the rules that govern our thinking or behavior, if necessary.

There is also the matter of accomplishment. Humans have sent remote systems out of our solar system. Robots have been to other planets, multiple times; there are three robots working on Mars right now. But only when a human steps on the surface of Mars for the first time will that be considered a momentous event "for all humankind."

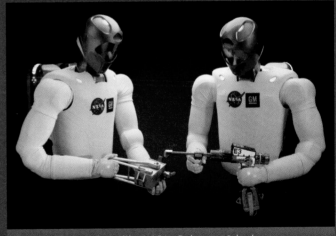

Figure 10-1 *Robots like NASAs Robonaut2 look very humanoid but are designed to work long hours in the harsh environment of space. They were specifically designed to use the same tools as their human astronaut **counterparts**.*

Courtesy of NASA

ORIGINS OF REMOTE SYSTEMS

You may think Unstaffed Aerial Vehicles or Uninhabited Aerial Vehicles (UAVs) are a relatively new invention. The UAV in some form or fashion has actually been around since 1916 when the Hewitt-Sperry *Flying Bomb* was developed, just 12 years after the first successful human flight. The Sperry UAV wasn't a true remote-controlled vehicle but paved the way for development of later UAVs. The only successful test flight occurred on March 6, 1918, when the *Flying Bomb* maintained steady and controlled flight for the preprogrammed 1,000 ft flight. Though the Sperry *Flying Bomb* ended in failure for its primary goals, key engineering breakthroughs occurred.

One of those breakthroughs was the technology for guiding the *Flying Bomb* based on **gyroscopic control**, which developed into the basis for autopilot control in today's modern aircraft (Figure 10-2). The Sperry *Flying Bomb,* once airborne, was designed to be completely autonomous. It followed a preprogrammed flight path, much like some modern-day UAVs with GPS. It had no real feedback loop other than the gyroscopic navigation system; where modern UAVs have an assortment of sensors to use.

Gyroscopic control:

to maintain control of a craft through means of a high rpm gyroscope. The principle of conservation of angular momentum keeps the rotating gyroscope steady while a control interface monitors the angle between the axis of rotation of the gyroscope and its housing.

Figure 10-2 A 1950s Honeywell Gyro Autopilot.

Courtesy of Honeywell International

The technology to make a remote flying craft is extremely complex. Unlike a ground-based or water-based vehicle, if a remote aircraft loses control, the usual immediate result is a crash. Think back to earlier chapters and all the forces and control surfaces that must be balanced to achieve straight and level flight. Now complicate this task by actually trying to fly to a target. Successful attempts to do this prior to the computer age were few and minimal.

One of the major benefits of a remote system is taking the human out of harm's way. Ironically, early in aerospace history, many remote systems were developed as weapons. But in recent history, remote systems have expanded to play a vital role in **ISR (Intelligence, Surveillance, Reconnaissance)** missions, exploration of space, and other missions that may be hazardous to humans.

ISR (Intelligence, Surveillance, Reconnaissance):

the prime directive for most military UAVs and remote vehicles to gather information through observation and monitoring a situation.

TYPES OF REMOTE SYSTEMS

As you learned in Chapter 2, remote systems have been a part of aerospace for more than 50 years. But what is the distinction between various remote systems? What makes one system autonomous? Is there such a thing as a completely autonomous system?

Remote-Operated Vehicle

A remote-operated vehicle is basically an extension of the human controlling the device (Figure 10-3). For example, if you've ever seen the Discovery Channel's *Mythbusters* show, you will see them frequently turn a regular car or boat into a remotely operated vehicle. You will also notice the Mythbusters are usually right behind the remote vehicle or maintain constant visual contact. In industry, remote-operated vehicles can have an array of onboard sensors, such as video, sound, or temperature, to help the remote pilot make informed decisions. Remote-operated vehicles must rely on their human pilots to make informed decisions regarding maneuvers and observation. In aerospace, a remote-operated vehicle is also called a **remote-piloted vehicle (RPV)**. The control station for an RPV can be as simple as a remote-control joystick or as complex as an entire room of computers and humans monitoring the **telemetry data** returned from the remote vehicle.

Telemetry data:
the measurement and reporting of data through wireless transmissions for use by a remote system operator to analyze data in real time or archived for later analysis.

Figure 10-3 An underwater ROV prepares for a dive. The tether provides power to the ROV and communication to the control ship.

© iStockphoto.com/Londai

Autonomous Vehicle System

Unlike an ROV or RPV, the **autonomous vehicle system (AVS)** can perform its designed tasks with almost no human interaction. AVS systems are gaining popularity in several disciplines from state police for accident investigation to utility companies for power line inspection, and other aerospace applications. Agencies such as the **Defense Advanced Research Projects Agency (DARPA)**

autonomous vehicle system (AVS):
a system that can operate, navigate, or perform tasks with little to no human intervention.

Figure 10-4 *The Boeing X45-A was one of the first stealth UAVs.*

Courtesy of NASA

have sponsored several autonomous vehicle challenges, through air, on land, and on or under water. A few notable DARPA projects are the Boeing x-45, Predator, and DARPA grand challenge (Figure 10-4).

Current autonomous vehicles require a human to begin the vehicle's tasks, and usually when the task is complete, humans are required to download data, refuel, or repair damaged or broken parts, for example. Today's large autonomous vehicles such as the Predator and Global Hawk can fly without any "pilot" interaction. After the pilot inputs the GPS destination or mission waypoints and instructions, the UAVs can take off, travel to the destination, return, and land safely without human interaction. It takes an enormous amount of sensor data, situational awareness, and communication to make an autonomous vehicle possible.

Hybrid Systems

When a mission spans several hours or humans are not required to have complete control for the entire mission, remote systems employ a combination of autonomous and remote-operated modes. **Supervised autonomy** is a type of remote system where a human operator can tell the vehicle where it should go, but the sensors on the vehicle determine the best way to get there. This mode is often used on very dangerous missions or instances where constant communication might be an issue.

Lockheed Martin developed a six-wheeled robotic vehicle called the Squad Mission Support System (SMSS) that can be remotely operated by a soldier via wireless video controller, put on autonomous mode to patrol an area, or follow the squad like a six-wheeled puppy (Figure 10-5). The supervised autonomy mode

Figure 10-5 **The Lockheed Martin SMSS. Notice the different sensors arrays on the front of the vehicle.**

Courtesy of Lockheed Martin

on the SMSS allows a soldier to tell the vehicle where it should go regardless of whether the soldier can see what is in front of the vehicle. The SMSS will avoid any obstacles in its path that look like a human or it thinks it cannot navigate. It can navigate around and return to its original prescribed path autonomously after avoiding the obstacle.

Another example of supervised autonomy is the Mars Pathfinder robots. The time delay between Earth and Mars can be anywhere from a few minutes to 22 minutes. That means it could take up to 40 minutes for a command to be sent and relayed back. Using supervised autonomy can alleviate some of the difficulty of navigating from such a distance. The Mars Exploration Rovers receive maneuvering instructions from a team of engineers on Earth who try to evaluate information relayed from the rovers before making any decisions on movement. After the engineers analyze and approve a plan, they send instructions to the rover. The rover tries to get to the new location and in the process avoids small obstacles that may impede its progress.

Even the best remote data isn't perfect. In May of 2009, while following commands to a new location, Spirit, one of the two rovers, became trapped in soft sand. NASA engineers spent almost 8 months trying to free the robot. In January of 2010, Spirit officially became a stationary research platform (Figures 10-6 and 10-7).

Figure 10-6 A 360° panoramic view of Mars from the rover Opportunity. The photo has been color enhanced to bring out details.

Figure 10-7 Checking soil samples turned up by the stuck wheels is one of the last missions of the sand-trapped rover Spirit. Notice the left wheel buried in the Martian sand.

US FIRST robotics competitions are perfect examples of hybrid remote systems designs. High school and middle school teams participate in various FIRST Robotics competitions throughout the country. The FIRST Robotics competitions introduce concepts in programming design and hardware design while reinforcing the concept of teamwork. Successful engineering design for FIRST Robotics requires proficient use of science, technology, engineering and mathematics (Figure 10-8).

Figure 10-8 US First Robotics is an excellent opportunity to learn teamwork skills and have fun with robotics.

Courtesy of NASA

COMMUNICATION SYSTEMS

Successfully operating a remote system of any kind without some kind of communication system would be quite difficult. The technological advances over the past few decades have brought improvements in signal quality, transmission distance, transmission speed, security, and method. Advances in traditional radio control and common technologies such as Ethernet, Wi-Fi, and Bluetooth have increased the availability of advanced remote systems in the hands of the average programmer or high school student.

Understanding Transmission and Reception

For any communication system to be successful, there must be a bare minimum of a transmit device and a receive device. A device could simply send a "ping" sound to another device, but you wouldn't know what that meant unless you previously established a **communication protocol** or method for translating the signal. For example, have you ever tried to use the wrong remote with a television? Your television does not know how to interpret the power command

Communication protocol:
description or rules for exchanging messages between systems. Protocols can be hardware based or software based or any combination of the two.

Figure 10-9 The OSI model for network communications. System 1 is sending information to System 2.

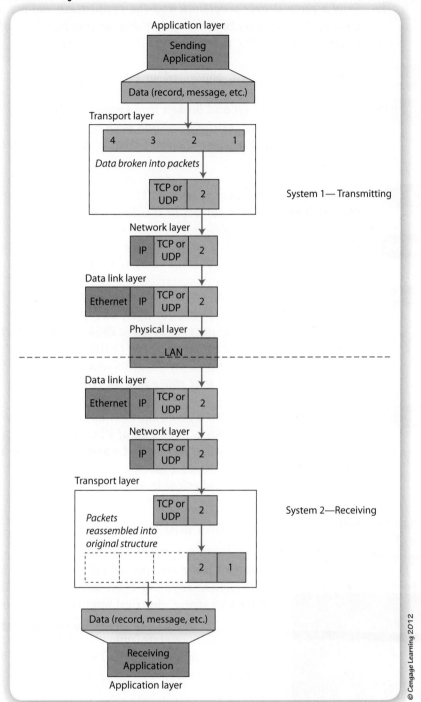

for your DVD player. A good example of communication protocol is the OSI model for networked devices shown in Figure 10-9. This works primarily for network communications but can be applied to robotics just the same (because many of these devices are now networked). The OSI model consists of seven layers starting with physical hardware at the bottom and the application (software) at the top.

How data is translated and transmitted by a system.

Step 1 ▸ *Send a request over a network and the application you are using creates data such as text, images, video, or commands. That application layer is the topmost layer of the OSI model.*

Step 2 ▸ *The data is presented, organized, or encrypted.*

Step 3 ▸ *Data is then lined up and broken into segments that the network can handle. At the front and back of each segment is information about where the packet data is going and where it came from.*

Step 4 ▸ *The transportation layer is responsible for making sure that routing information is correct or available.*

Step 5 ▸ *At this point, the data is routed on the network layer in packets, following the path determined by the routing devices.*

Step 6 ▸ *The data link layer is the physical address of the device (MAC address) and is usually unique for every device. From there, the network layer reads the packet. Because devices are connected to a network, they will usually tell the routing device what the MAC address is and pass this information along so the transport layer can determine the best route and location.*

Step 7 ▸ *In the physical or hardware level, all information is transmitted as bits, which are the basic 1s and 0s of binary communication. This is also the level that includes the transmission media and signal strength. (When troubleshooting a communication issue, you usually start at the physical level.)*

Figure 10-10 Fiber optic bundles can carry hundreds of strands in a very tight pack. Each one of those strands can carry hundreds of signals working on different light frequencies.

©iStockphoto.com/MsLightbox

Mediums and Wireless

Most communications systems can be categorized into two basic systems: wired and wireless. Wired can include anything from sound waves traveling through a string, electrical impulses in a metal wire, or light waves traveling on an optical fiber. Wireless transmissions operate "unguided" by broadcast in all directions (omnidirectional) or in directed beams (unidirectional).

Wired Communications

Even though wired transmission is not in a vacuum, the transmission speed over a wire is about 66% that of the speed of light or nearly 200,000 km/s, so it is still much faster than sound.

Today most modern long-distance communications occur over fiber optic cables (Figure 10-10). Fiber optic can carry a light wave thousands of miles over a very thin strand (about the size of a hair). Improvements to the transmission and receiving devices of fiber enable multiple data streams to travel over the same data fiber simply by using different light frequencies. Some advantages of using fiber optics include that light waves travel slightly faster than electromagnetic waves, and the improved signal to noise ratio possible when fiber optic strands are bundled into multistrand cables because the signal in one fiber does not cause interference in neighboring fibers. However, the cost of fiber still remains higher than metallic wires, and copper and aluminum still remain the ideal choice for electricity and power transmission. Copper-based wired transmission is ideal for short distances or when flexibility or sharp bends in the communication cable may occur.

Wireless Communications

You can't see it, but every day, we are bombarded with all kinds of wireless electromagnetic transmissions (Figure 10-11). Think about all the signals traveling through the air right now—TV, AM/FM radio, satellite cable, cell phones, wireless networks, cordless phones, GPS, and the list goes on.

Figure 10-11 A typical communications tower high atop a mountain or cliff. The higher the tower, the further the signal can carry.

©iStockphoto.com/bluebird13

Fun Fact

New technologies promise to give us the ability to harvest the energy from all the wireless signals in the air to recharge small batteries and to operate small electronic devices. Companies are beginning to discuss products such as RCA's Airpower device that are intended to replace the battery pack in traditional cell phones, iPods, and PDAs. Does the technology work? No products have yet been released, so only time will tell.

You might consider sound a type of wireless transmission. However, sound does need a medium to travel through. That medium can be a gas, a fluid, or a solid. The denser the material, the faster the sound wave travels. Wireless electromagnetic transmissions (EM transmissions) do not require a medium, making them a great option for space communications, and they are much faster than sound.

Fun Fact

Wireless transmissions actually travel faster without a medium present. EM transmissions travel at the speed of light (nearly 300,000 km/s in a vacuum). Sound only travels at 0.343 km/s (sea level at 20 °C). Because of the huge difference in speed, you see the flash of lightning long before you hear the sound. You can approximately calculate the distance from the lightning strike by counting the number of seconds between flash and thunder and dividing by 5. This gives you approximate miles (divide by 3 for kilometers). For example, if 10 seconds pass between lightning flash and thunder, the strike was about 2 miles away.

As with wired communication, wireless still needs a transmission protocol and the ability to interpret signals on different frequencies. Wireless signals can be more prone to interference from weather conditions than wired, and security is an issue as well. Because wireless is generally broadcast (there are some cases of point-to-point wireless), security is a big issue. Signals are often encrypted and decrypted, which consumes processing time for the signal.

Your Turn

Check Your Understanding

In your notebook, take a quick moment to try these calculations. Watch for units.

1. Ignoring delays with routers and other devices, how long would it take a signal to travel from computer to computer, over wire, 15,000 km apart?
2. Assuming the speed of light is 280,000 km/s, how long would it take a wireless signal from Earth to reach another station on Earth via a geo-synchronized satellite at 36,000 km above Earth?
3. How long would it take a signal to reach Mars when it is at its closest orbit of 36 million miles? What about its farthest distance of 250 million miles?

Why is all this important for remote systems? The closer in proximity devices are, the closer to real-time the transmissions get. The farther out our remote system, the more we have to account for communication delays. Remember, all these transmission times we have discussed so far are purely travel time without any delays caused by routers, processing, encoding/decoding, encrypting/decrypting, and other signal handling steps (Figure 10-12).

Figure 10-12 *The Endurance is an autonomous underwater vehicle (AUV) designed to work in frozen bodies of water.*

Courtesy of NASA

The main characteristics of *wired* communication include the following:

▶ Less prone to interference

▶ Somewhat secure

▶ Guided point-to-point, not broadcast

▶ Controllable through routing

▶ Must lay pathways before communications

▶ Distance limited

▶ No space transmissions

▶ Subject to shorting or failures

▶ One point of failure if cut or broken

The main characteristics of *wireless* communication include the following:

▶ Wide broadcast

▶ Ideal for space communications

▶ Prone to interference from other man-made or natural radio sources

▶ Can be "secured"

▶ No need to pull cables or wires to remote locations

▶ No tether to follow or get tangled

▶ Does not travel well through dense mediums

▶ Can cover great distances without additional weight

You may wonder why all communications hasn't become wireless. The problem is that only a limited number of frequencies are available to use for transmissions that are considered safe. The electromagnetic spectrum covers a wide range of frequencies, many of them harmful to humans (Figure 10-13). In the next section, we will take a look at how those radio frequencies work and how we send information over those frequencies.

Figure 10-13 Visible light is a very small part of the entire electromagnetic spectrum. The longer the wavelengths, the less harmful to humans.

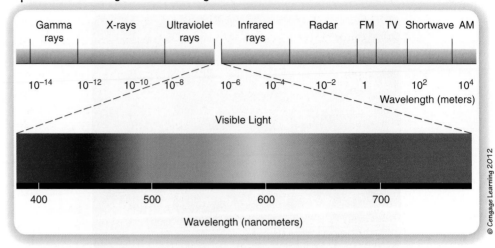

Understanding Radio Transmission and Reception

The more promising technology that led to the success of the modern UAV rested in radio control (RC). Nikola Tesla is credited with using the first radio-controlled device in 1898 on a small boat, but the credit for the first radio-controlled aircraft belongs to Archibald Low. Low developed, with the help of the Royal Flying Corps, the "aerial target (AT)," which, for strategic purposes, was a misnomer because the AT was really a flying bomb designed to carry a warhead. On March 21, 1917, the RC aircraft was born with a successful test flight.

Unfortunately, Low's ideas were years ahead of their time and did not reach their full potential until the late 1930s and World War II. Several RC projects from the Axis and the Allies were developed primarily using **electromechanical** devices to operate relays to control actuators. This system was bulky and prone to malfunction but was used up until the 1960s when the introduction of **solid-state** control systems improved radio control. The basic mechanism was still the same: A signal from the ground was received by the radio control receiver, which transmitted to a relay, which in turn moved a control actuator and a control surface. Though modern RC equipment has much smaller components, providing more control channels, elements of both radio and gyroscopic control remain in today's autonomous vehicles with visual and sensor telemetry data feeds back to the pilot on the ground (Figure 10-14).

Solid state:

most modern devices are solid-state devices wherein the electrons are contained within solid materials, not gas tubes or relays. Commonly associated with printed circuit boards (PCB) and increasing miniaturization of the technology.

Figure 10-14 Radio Control components continually get smaller and lighter as technology improves. This helicopter works on the 2.4 GHz radio range.

In the previous section, we looked at the differences between wired and wireless communications and the differences between light and sound waves. There are two fundamental types of waves: **longitudinal** and **transverse**. Sound waves are generally longitudinal; electromagnetic is transverse. Regardless of the type or medium that is traveling, all waves can be described using their period, frequency, and amplitude. As a wave passes a given point or particle, that particle is displaced; the height above that particle's equilibrium position is called the wave amplitude. The distance between successive peaks or valleys is called wavelength. A single wavelength is the distance the wave moves forward during the time the particle repeats a given cycle. As a dimension of time, this cycle is described as the period or time between waves.

Example

As an early navigation aide, pilots and ship captains figured out their location by knowing the frequency that a certain light tower rotated. Pulsars in space, which are rotating neutron stars that emit a beam of radiation, are also described by their period, or how much time passes between flashes.

To describe how often the cycles happen in a given time, we take the inverse of the period (T) and call it frequency. When comparing the number of cycles in a second, we use Hertz.

We often use the prefixes Kilohertz (1000 Hz), Megahertz (1,000,000 Hz) and Gigahertz (1,000,000,000 Hz) to describe very high frequencies.

Figure 10-15 *The frequency pegboard is one of the first things visible on an RC flying field.*

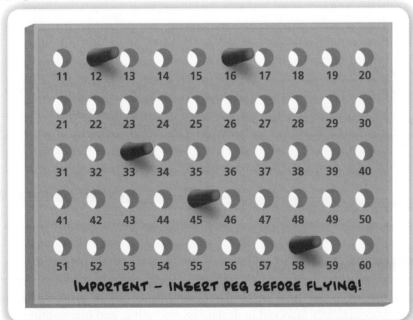

© Cengage Learning 2012

Digital versus Analog

Data transmitted over wires or wireless can be categorized into **digital signals** or **analog signals**. Recently, in the United States, all analog television signals were turned off in an effort to make room for other uses on the spectrum. All that remains of broadcast television is a digital signal that only newer TVs or special adapter boxes can pick up. Digital radio is increasing in popularity, and it too will

GRASS STRIP ADVENTURE

Certain wireless devices are set to communicate only on specific frequencies of the electromagnetic spectrum. If you get a chance to check out your local flying field, take a look at the radio frequency board (Figure 10-15). You will see a peg system or sometimes the RC pilot's Academy of Model Aeronautics (AMA) ID card placed in the slot for a particular channel. This helps prevent frequency interference on the flying field. Typical model RC aircraft operate on the 72 MHz frequency with 50 channels (frequencies from 72.010–72.990 MHz in 0.02 MHz increments). Other common frequencies are also used for RC boats and cars in the 27 MHz, 50 MHz and 75 MHz bands. The new technology on the radio scene is the digital spread spectrum radio that operates on the 2.4 GHz range. These new radios are digitally paired between transmitter and receiver, which are designed to prevent channel interference.

Digital Spread Spectrum (DSS) technology is based on the idea that the transmitter and a receiver can be "bound" to each other so that they share a common set and sequence of frequencies through which they both "hop" during signal transmission. The transmitter breaks its signal into parts that are each transmitted on a different frequency, thus reducing the average signal strength on any particular frequency significantly. The receiver puts the entire signal back together by collecting from all the different frequencies in the same sequence it was broadcast in. This approach makes the signal much more secure, harder to intercept, harder to jam, and able to overcome many problems with interference.

probably completely replace analog radio. What does this mean to the signal? Basically, an analog signal has a continually variable signal.

If you were to look at an analog signal on an oscilloscope, you would see any type of sine wave, triangle wave, or sawtooth type wave. A digital signal on an oscilloscope is typically a square wave with only two states: on (1) or off (2). Noise—interference on the signal—can affect both analog and digital signals. Digital tends to be less affected because the digital processor ignores any value that isn't quite a 1 or a 0, but too much noise can completely kill a digital signal. In analog systems, that noise becomes part of the signal and can be difficult to filter out. As communication technology continues to improve, digital signal transmission is becoming simpler and more economical than analog transmission (Figure 10-16).

Figure 10-16 *A comparison between a typical sine wave (analog) and a square wave (digital).*

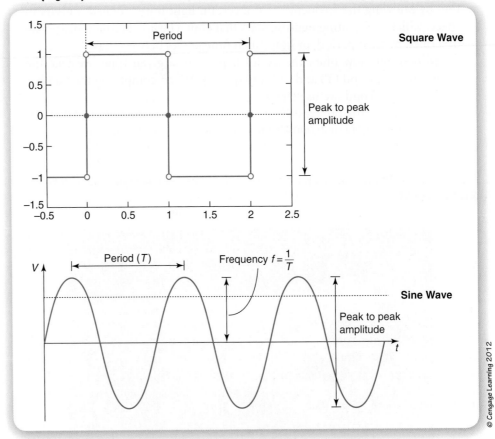

SYSTEMS ENGINEERING

Because most of this chapter is devoted to talking about remote systems, we should consider the complexity involved in designing, operating, and maintaining remote systems. Systems engineers usually evaluate the design of such complex systems. In essence, systems engineering deals with identifying system goals, designs, constraints, feature importance, tradeoffs, risk management, and, ultimately, the best design for the customer and task needs.

Let's look again at the Global Hawk UAV from a systems engineering point of view (Figure 10-17). The customer (NASA, military) needed a method of observing remote areas on a moment's notice. Previous technologies such as satellite or

Figure 10-17 *The Global Hawk UAV has a long thin wing much like a sport glider or U2.*

Courtesy of NASA

U2 had both advantages and limitations. To improve beyond these existing systems, the Global Hawk needed to be able to loiter over an area for extended periods of time, have a small radar signature, and carry a large array of sensors and cameras to relay information back to base. It also needed to be completely autonomous or piloted, be transportable to anywhere in the world, and be able to avoid enemy attacks. At some point in its development, a group of engineers discussed the goals and features of these aircraft members of the military. The engineers gathered information such as feature importance, constraints, and ultimate goals before proceeding with a design. As part of the systems engineers jobs, they also helped develop the communication network, maintenance recommendations, testing procedures, and many other aspects related to daily operation of the Global Hawk.

Support Systems

As mentioned earlier, there are three robots on the surface of Mars performing various tasks. Two of the robots, Spirit and Opportunity, began their journey on Mars in January of 2004 (Figure 10-18). The rovers were originally designed to last only 90 days on the Martian surface, but as of September 2010, the rovers had logged more than 30,804 meters combined. That's almost 20 miles and 6 years past their mission objective! These robots have been so successful for two reasons:

▶ **Innovative hardware/adaptable programming.** Although both robots have had issues along their journey, they are still able to adapt to changes. Redundancy in the hardware of the robots systems, such as the six independent drive wheels arranged in a **rocker-boogie** configuration, helped keep the rovers mobile even when one of Spirit's wheels failed early on (Figure 10-19).

▶ **Battery charging/protection.** Large solar arrays charge the batteries, and radioisotope heaters keep the batteries warm in the harsh Martian environment. Several other precautions are employed to prevent dust from entering the electronics and actuator components.

On the software side, the Mars rovers are considered to be supervised autonomous vehicles. Humans might tell the rover where to go, but it is up to the rover to figure out the best path to get there. The robots use a combination of 3D mapping to select

Figure 10-18 Artist's rendering of the Mars Exploration Rovers. The rovers have logged a total of more than 16 miles on the Mars surface.

Courtesy of NASA

Figure 10-19 The 1997 Pathfinder rover employed an early version of a rocker-bogie suspension. The suspension has one attachment point to the chassis and is a series of levers.

Courtesy of Jasen Ritter

the path of least resistance. If the robots had to wait for human input for every move, their progress would be painfully slow. Remember the time delay; you can imagine the problems caused by having to wait up to 40 minutes from the time you told the robot to turn until receiving visual or telemetry information that it was successful!

Programming

Most successful programming designs are adaptable; they rely on sensor information and the program itself can analyze variables. Early in the computer age, programming a set of commands required making punch cards. A simple program to do basic calculations used hundreds of these cards! Now programmers have a plethora of options for programming languages, programming interfaces, and even GUI-based graphical programming that flows like a block diagram. A novice programmer can simply place and connect blocks.

Fun Fact

If you wanted to get involved in robotics when it was a relatively new technology, you usually had to be involved with a university or government installation. Now the popularity of robotics has grown so that almost anyone can get involved. In fact, the age requirement for the Lego Mindstorms NXT robotics kit is age 8 and up! Visual programming languages have made it much easier to master the basics of programming a robot.

When first starting a program, often engineers and programmers use techniques such as pseudo-code or human language programming to describe or outline what they want the program to do.

For example, think about retrieving a beverage:

▶ Detect level of thirst.

▶ If thirsty, then initiate beverage retrieval; otherwise, reevaluate level of thirst in near future.

▶ Initiate beverage retrieval.

▶ Stand up.

▶ Access environmental status map to determine general direction to beverage's current location.

▶ Turn to face general direction of beverage.

▶ Avoid obstacles while advancing . . . and so on.

Later, they go back and fill in the details in the proper programming language context (Figure 10-20). To go with all these programming languages are a multitude of platforms to help the novice robotics explorer, including anything from easily changeable block-based platforms to custom built specialty parts and controllers.

Your Turn

You Be the Robot
You can see what it's like to be a robot with a little human experiment. List all of the sensors you, as a human, have available and how those senses could get you through a simple maze. Now how difficult would it be if you were only left with one or two senses, such as touch and balance (yes, balance is a sense). If you were blindfolded with noise canceling headphones on or in a very quiet room, how would you "program" yourself to get out of the maze? Think in terms of creating an algorithm to follow, and then try it out.

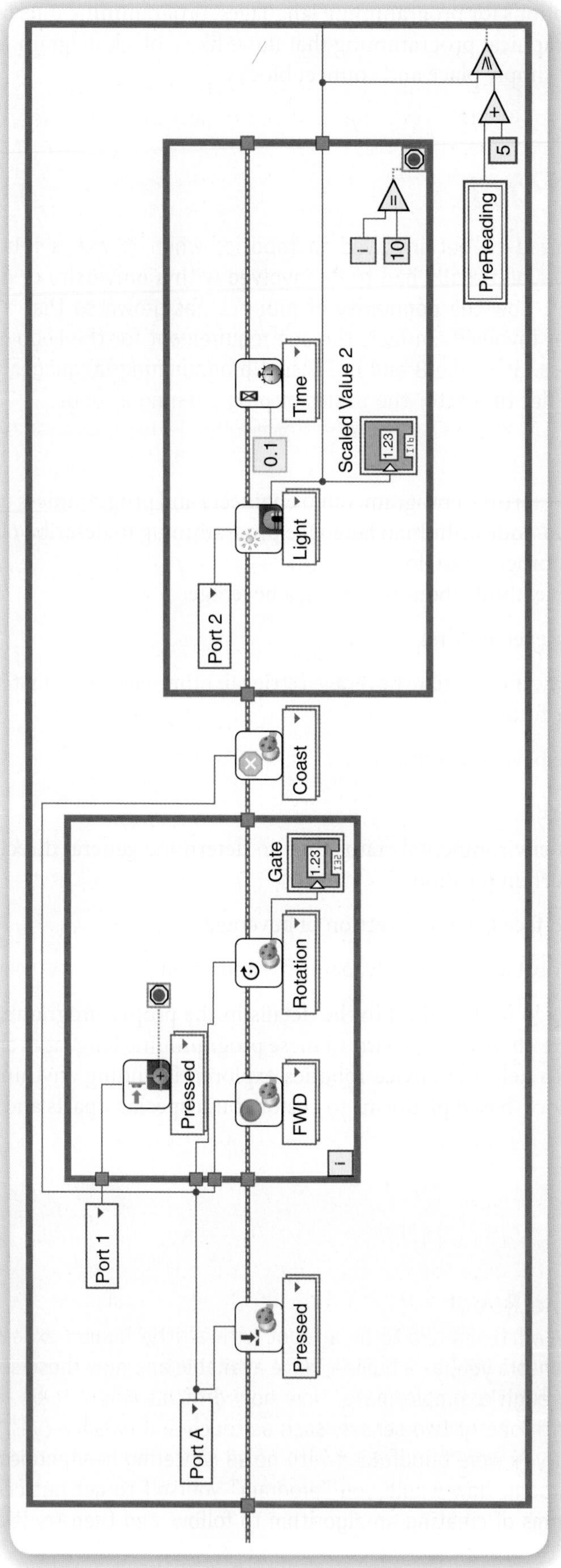

Figure 10-20 Sample graphical programming within Labview.

Artificial Intelligence

When you hear the term **artificial intelligence (AI)**, you might think of science fiction robots that can think, act, or look like humans, such as the self-aware robots in the science fiction collection *I, Robot*. The reality is that remote systems of today already have demonstrated AI. That is, these robots can perceive their environment and make decisions based on that environment.

Faster processing, more efficient programming languages, and a diversity of sensors make AI possible. In the early days of processors, they could only do a few hundred or thousand operations a second. Just 20 years ago, real-time 3D image processing at any decent vehicle speeds would have been almost impossible (Figure 10-21). With modern processors, robots now have one to eight processors, or more, working together, capable of handling billions of operations per second. These processors are able to handle 3D image analysis and monitor arrays of other sensors and data at the same time, in real-time (meaning as soon as the sensor detects information). With more processes available, programs are more complex, performing sensor readings and mathematical computations instantaneously.

artificial intelligence (AI): design of a system that can perceive its environment and take actions to maximize success for a particular task or set of tasks.

Figure 10-21 *Stanford University's 2007 DARPA Challenge entry. In 2006, none of the teams completed the course. In 2007, three teams successful completed the course.*

Stringer/AFP/Getty Images

Instead of hard coding variables and waiting for other instructions to compete, programmers can monitor sensors and telemetry with adaptable programming. The Mars rovers, Spirit and Opportunity, use their 3D image-mapping system in conjunction with rotational sensors in the wheels to compensate for slippage on the terrain. If the 3D image has not changed significantly or equal to the amount of rotation the wheels have turned, the robots know they have slipped in movement and compensate. A simple adaptable program might be one that uses math formulas for variables instead of hard-coded numbers. To solve this problem, the line-following robot you build in an engineering class can look for a change in light sensor values instead of referring to values predicted in advance of the robots activation. Especially when the robot operates in a setting that is inaccessible (such as the surface of Mars) prior to the mission launch, this allows the robot to be much more adaptable than a robot that only compares its navigational readings to hard-coded estimated light values.

I, Robot and Strong AI

The eventual outcome of artificial intelligence will be "strong AI" where robots can learn, reason, and plan—basically, think. Robots are very useful tools, but caution must be taken as technology is pushed forward. When factory robots were first introduced in automobile manufacturing plants, it created quite a controversy because humans jobs were replaced by machines. Imagine what might happen as robots replace even more daily human activities (Figure 10-22).

Figure 10-22 *Robots like this lawn-mowing robot are becoming commonplace. What kind of sensors might this robot use to complete its tasks?*

Courtesy of Ames F. Tiedeman

REMOTE SYSTEMS HARDWARE

Building remote systems out of premade kits is definitely possible. Lego, Parallax, VEX, and several other companies have highly adaptable, expandable robotics platforms useful to both the novice and expert builder. Although the many platforms may have different pieces or ways of connecting pieces, they all generally have the same components. As specialization of a remote system increases, so do the specialty pieces. You can't build a Global Hawk UAV with off-the-shelf components, but with a little creativity, you could get pretty close.

Input Devices

The primary goal of many remote systems is to provide feedback from a remote location. Without ways to gather information, the remote system is basically a remote toy and can quickly become useless once out of direct visual contact with its operator. The list of input devices employed on remote systems is nearly limitless, however, most remotes systems carry a specific package of sensors that enable the systems to complete the vehicle's primary and secondary objectives.

Some examples of sensors used by a remote system for positional awareness include the following:

► Video

► GPS sensor

► Three-axis rotational sensor or gyroscopes

► Three-axis accelerometers

► Sonar or RADAR (Radio Detection and Ranging)

► LIDAR (Light Detection and Ranging) using lasers

► Additional rotation sensors and monitors attached to actuators, engines, thrusters, or other onboard devices

Payload sensors, which are sensors required to do specific mission tasks, include the following:

► Video

► Infrared video

► RADAR

► Atmospheric sensors

► Soil/rock sample testers

► Gas spectrometers

► Electromagnetic detectors

Output Devices

The remote system can observe its surroundings by using input sensors. But to react or move, the system needs output devices. In open-loop control systems, there is no feedback from sensors to control the behavior of the system. As you can imagine, this can cause problems for a remote system. A closed-loop control system is capable of making decisions based on sensor input and is the preferred method for most remote systems. Take for instance a GPS receiver, accelerometer, and rotation sensor attached to a microprocessor interfaced to control surfaces. The microprocessor can evaluate the planned GPS route to its actual route. If it needs to make adjustments, the microprocessor commands the control surface actuators to move in the desired direction, while monitoring the accelerometer and rotational sensors to keep the system stable.

The following is a sample of output devices:

► Various propulsion systems

  ► Engines

  ► Electric motors

  ► Thrusters

  ► Pumps

► Actuators

► Sound- or light-producing devices

► Weight-shifting devices

► Data streams

PUTTING IT ALL TOGETHER

As you have learned, a UAV is defined as a remote controlled or fully autonomous vehicle that is intended to be reused many times. Today's UAVs come in many shapes and sizes, all with distinct purposes. Some uninhabited aircraft work independently or are part of a system known as an Uninhabited Aerial Vehicle System or just Uninhabited Aerial System (UAS). The attack-capable Predator, which has flown many hours over the Middle East and Bosnia, is a system that includes up to 4 aircraft with sensors, a ground control station, a satellite link, and approximately 55 personnel, and is deployable anywhere in the world. Attack-capable UAVs such as the Predator are also known as Uninhabited Combat Aerial Vehicles (UCAV).

The idea of using UAVs for reconnaissance was a product of the cold war, in which both the United States and the Soviet Union developed and deployed long-range reconnaissance drones in secrecy. The primary role of the Global Hawk UAV is reconnaissance. Its huge wing span and high aspect ratio is similar to what is found on sport gliders. This efficient design enables the Global Hawk to have a range of more than 14,000 nautical miles and a flight duration of up to 42 hours.

Current Remote Systems

Remote systems take on a variety of roles and designs. As we discussed, specific missions call for specific hardware and capabilities. Current UAVs can perform multiple roles. Following the success of the Predator UAS, UCAVs and UAVs continue to grow in many capabilities. Some experts believe the F-35 JSF will be one of the last manned fighters built, being replaced by smaller, cheaper, and more maneuverable UCAVs. Remember, current fighter aircraft have much greater power and maneuverability than can be endured by the pilot but are governed by computer control to not exceed those limits. Additionally, cabin space, life support, and safety systems add precious weight to the aircraft. Moving the pilot to ground-based systems addresses those issues and takes the pilot out of harm's way.

At the same time, this raises other issues involved with UAVs, especially those designed for attacking targets (Figure 10-23). Leaving the pilot on the ground to watch through a camera could impact their decisions, not allowing the pilot to have full understanding of a situation or target. And still more concerning will be attack decisions completely made by a computer. That scenario may still be a long way off, but the evolution toward that goal is progressing.

Just as some UAVs and UCAVs are designed for stealth capabilities, other UAVs are designed for portability. The Army and Marines use a Micro Unmanned Aerial Vehicle (MAV), called the Raven, which is small enough to fit in two packs that weigh a combined total of 10 lbs (4.3 Kg) and can be launched with a two-man crew. The Raven provides Special Forces teams with day or night surveillance of a target area to allow evaluation of possible hazards before entering an area. MAVs like the Raven, which can be directly flown by joystick or a GPS-plotted course, have been employed by the military since 1989.

Figure 10-23 NASAs version of the non-combat Predator UCAV.

Courtesy of NASA

Fun Fact

Helicopters have also been employed as UAVs. They have the unique advantage of hovering in a stationary position, which can be very useful for a small UAV to monitor a target area. However, this poses two problems: A stationary target is an easy target, and rotary wing aircraft can be very loud, making stealth difficult. A more entertaining use for UAV helicopters is for flyby scenes for movies. A ground based operator equipped with a large RC helicopter and movie camera can be much more affordable than renting a real helicopter for the same scene. Many large police forces now use RC-driven and GPS-driven helicopters to assist in the evaluation and analysis of car accident scenes (see Figure 10-24).

Figure 10-24 The turbine engine ChopperCam allows for steady camera shots and has been used for companies like Disney, Paramount and others.

Future of Remote Systems

Computer technology continues to improve at exponential rates. Sensors, actuators, and processors are being developed on the nano scale so that UAVs the size of insects are real possibilities. Swarms of individual micro remote systems could gather and share collective information acting as one large system. Similar but not the same as swarm behavior is the concept of **Systems of Systems (SoS)**, which is a hot topic in remote system design. SoS design involves taking multiple independent technologies and merging the collective resources into a larger meta-system.

Using the Global Hawk UAV as an example, there is much more to that system than just the UAV. The Global Hawk is tied into satellites that relay communication information from above as well as communicating with groups on the ground. Operators of the Global Hawk can observe weather patterns from other satellite systems and incorporate that data into communication issues they might encounter with the UAV (Figure 10-25).

Figure 10-25 Engineers watch over flight data in a control center for one of NASAs Global Hawk UAVs.

Ground troops in the area can contribute or use information from the system as an "eye in the sky." The same UAV can be communicating information back to aircraft carriers to help coordinate rescue or strike efforts. SoS designs tend to be on the extreme in terms of complexity. Tools such as Analytical Graphics, Inc. (AGI) Satellite Tool Kit can help engineers speed up the development of relationships between systems.

Remote Systems Competitions

Nothing motivates design changes and new technologies better than a little competitive spirit and sometimes prize money. The Association for Unmanned Vehicle Systems International (AUVSI) hosts competitions for universities to develop cheaper and more reliable UAVs. The contest parameters require teams to do a complete

autonomous mission, including autonomous takeoff, GPS navigation, automatic target identification, and autonomous landing. Every few years, the association changes the contest to keep challenges competitive based on current technology (Figure 10-26).

Figure 10-26 The University of Texas Arlington Autonomous Vehicle Labs placed 3rd overall in the 2008 AUVSI Challenge using several off-the-shelf components to save costs.

Courtesy of Dr. Arthur Reyes

Fun Fact

Get Involved with Robotics

The Google Lunar X prize (www.googlelunarxprize .org/) core components involve launching a rover to the Moon and sending back HD video and images. Google is offering bonuses for extra tasks such as traveling away from the landing site or finding NASA or Russian space hardware left behind on the Moon's surface.

The Moonbots competition (www.moonbots.org/) is a co-op between Lego and Google to bring similar lunar mission objectives to a smaller scale.

US First (www.usfirst.org/) robotics has many levels of competition for all ages from kindergarten to high school. The First Challenges cover a wide variety of objectives, but all of them include teamwork and innovation. No matter what, try to find a competition or make your own within your school. Troubleshooting hardware and software issues with robotics is a great problem-solving tool.

More and more, robots have become a daily part of our life—from iRobot™ Roombas™, robots that clean gutters, cars that can park themselves, and military robots that report tactical information back to commanders. The challenge to humanity is to develop robotic technology that enhances our quality of life without becoming reliant on its capabilities. Remote systems are here to stay and will be more and more present in our daily lives. Although these systems protect human lives, make dangerous exploration possible, and make our daily life more convenient, we must remember that this technology is a tool to be used and understood, not simply to do the work for us. Cars that park themselves or slow down automatically when you get too close to a vehicle are great safety features. These are features that are nice to have for that one moment when, we are distracted by something else. If we rely on these technologies full time, we risk losing our ability to do even the simplest tasks later on (think about how much you use a calculator for math).

Careers in Aerospace Engineering

A WOMAN WITH A MISSION

For a lesson in how to turn your dream job into reality, spend a little time with Tracy Van Houten. Even as a teenager, Van Houten knew she wanted to work for NASA's Jet Propulsion Laboratory, the research-and-development wing where unmanned missions are continually being launched to explore our solar system and beyond. Unlike so many dreamers, however, Van Houten actually did something about her dream. In addition to picking up a couple of very useful degrees—a bachelor's from California Polytechnic State University, San Luis Obispo, and a masters from the University of Southern California—she networked. That's right, she networked, starting when she was 15 years old.

"Once I knew that JPL was the place I wanted to be," Van Houten says, "I started making contacts. And anytime I met someone even remotely connected with JPL, I would ask for his or her business card. By the time I graduated, I had this huge binder full of all the contacts I'd made there. I probably sent out 50 resumes to people on my contact list. I got interviews for three different jobs at JPL, and job offers from all three of them. Basically, I got to pick the job I wanted."

On the Job

Today, that job involves Jupiter—actually, Europa, one of Jupiter's moons. Van Houten is a member of the team putting together the Europa Jupiter System Mission. Still in the formulation stage, this interplanetary mission, which is baselining a launch in 2020, would send a spacecraft to orbit Europa, looking for signs of life.

"Europa is one of the icy moons," Van Houten says, "and there's believed to be a liquid water ocean under the icy surface, which means there's the potential for life to have formed. We would be looking for the building blocks of life. We would also be confirming the existence of that ocean and characterizing the ice shell. How thick is it? How does it interact with the ocean below?"

Tracy Van Houten

Working as a flight system systems engineer, Van Houten is helping design the spacecraft that would carry the scientific instruments across the solar system. "The biggest technical challenge is Jupiter's high level of radiation," she says. "Electronics don't do well with radiation."

An additional challenge for Van Houten personally is that she's coordinating the work of various subsystems, which means getting everything and everybody to work together. "It's a people-interaction job," she says. "I like being involved in the big picture."

Education

Sophomore year of high school was when the light bulb went on over Van Houten's head. "I took this pre-engineering and design course, where I was one of only two or three girls in a class of 40 students," she says, "and I got the highest grade in the class. I found out I was really good at designing and analyzing."

Cal Poly turned out to be a great place to apply those skills. "The philosophy there is 'Learn by Doing'," Van Houten says. "I loved getting all that hands-on, real-life experience while still in school."

She also became involved with the Society of Women Engineers, serving as president of the Cal Poly section, and she remains active in the group today. "We focus on networking, professional development, that kind of thing," she says. "The Society is celebrating its 60th anniversary, and I'm sorry to say that the numbers still aren't that good. Today, the percentage of women engineers is in the low teens."

Advice to Students

In addition to putting together your own business-card binder, Van Houten recommends you embrace any leadership opportunities that come your way, especially if you're female. "A lot of girls who are good at math and science aren't steered towards engineering," she says. "They're told to become math or science teachers. Don't allow people to do that to you. Women who drop out of engineering tend to have much higher GPAs than men who drop out. Don't drop out. Keep going."

SUMMARY

- Remote systems encompass several categories that are mainly defined by the level of human interaction. Remote systems can be direct extensions of human input or completely autonomous.

- Without an effective communications design, remote systems would not be possible. Communication designs can be guided through wires, optical cable, or another medium.

- Wireless communication is broadcast information that can be received as long as the receiver is in range.

- Most remote systems, like aerospace vehicles, are very specific and deliberate in design function or purpose.

- In complicated remote systems, the support team to operate a remote system can be more expensive and complicated than a human-onboard system.

- Pseudo-coding and human language coding are methods of outlining a program before actually coding in a specific language.

- Input devices are sensors, timers, or other signals that send information about the environment to the remote system. Using sensors in a program is known as closed-loop programming. Programming that does not receive information from the environment is known as open-loop programming.

- Programming languages provide communication for input and output devices and software interfaces.

- Separate systems interacting with one another in an SoS design result in a meta-system that can create a greater span of usefulness for the remote system. The whole system becomes greater than the sum of the parts.

- Competitions inspire new technologies, develop problem solving skills, and offer a wide variety of challenges to students.

BRING IT HOME

1. What are the basic components of a remote system?
2. What do you need to properly communicate with a system?
3. Using the Internet or other resources, find a remote vehicle or device that was pivotal in the development of remote systems.
4. Create a matrix comparing and contrasting the use of remote operated control, autonomous control, or supervised autonomy.

5. Give a presentation on how you could use a UAV in your daily activities. How could it improve your life? How could it hinder your life?
6. Make a list of the electronic systems you use daily that are based on the sensor, processing, and output cycle for product automation. Try to live a day or two without these technologies.

EXTRA MILE

1. Encourage your instructor to help you find a local or national robotics competition.
2. Start a robotics club.
3. Find an RC club to learn more about proper radio control procedures.
4. Talk to representatives of a local university to see if the university is involved in any robotics competitions. Find out if the university has any demonstrations or how you might get involved.

5. Check out the AUVSI video of incredible unmanned vehicles at www.youtube.com/watch?v=C62JSgJo39E&feature=related.

GLOSSARY

2 Gs: Roll an airplane into a 60° banked turn, and the total gravity force felt by the pilot is twice the force of gravity (known as 2 Gs).

A

Acceleration: the rate of change in velocity.

Acceleration, Centripetal: a change in the velocity of an object due to a force acting directly toward the center point of rotation.

Actuator: a device designed to create a mechanical force that causes something to move. Actuators can open or close valves or move a control surface to a position.

Adverse Yaw: the tendency for the nose of the aircraft to move in the opposite direction of a turn due the increased induced drag on the rising wing and reduced induced drag on the descending wing.

Aerodynamics: the study of the effect of air flowing over and interacting with objects and the production of forces such as lift and drag.

Aerospace Engineer: a professional trained to design, test, and evaluate craft that move through the atmosphere or the vacuum of space.

Aerospace Engineering: the discipline of applying scientific, technological, and mathematical knowledge to create solutions to problems related to operating within and beyond the atmosphere.

Afterburner: a device for injecting addition fuel into the exhaust stage of a gas turbine (turbojet) engine to drastically increase the total thrust it produces.

Aileron: a movable control surface attached to the trailing edge of the wing. Deflection of the ailerons is in opposite directions on the left and right wings.

Airspeed Indicator: a Pitot-Static instrument that measures and communicates to the pilot the relative speed of the air and the vehicle. Note that this is commonly different from the ground speed of the vehicle due to wind.

Alloys: a material made out of a uniform mixture of various metals.

Almanac Data: allows the satellite to transmit the orbital characteristics for every satellite as well as corrections for signal degradation due to the ionosphere and time encoding down to the GPS receiver.

Altimeter: a Pitot-Static instrument that measures the aircraft's height of flight compared to a common altitude of mean sea level (MSL).

Analog Signals: data streams interpreted by voltage levels or other similar means. Can be any value representation.

Angle of Attack (AOA): the measure in degrees between the chord line of an airfoil and the relative wind.

Aphelion: the point of farthest approach between an orbiting body and the Sun.

Apogee: the point of farthest approach between an orbiting body and Earth.

Artificial Intelligence (AI): design of a system that can perceive its environment and take actions to maximize success for a particular task or a set of tasks.

Ascending Node: for planets with a tilt of their axis, relative to the perpendicular to the plane of their orbit, the object that they are orbiting will appear to cross their celestial equator twice per orbit. The two locations where this occurs are called nodes; the ascending node is where the object moves from below the equator to above the equator.

Aspect Ratio: the ratio between the average chord of the wing and its span from wingtip to wingtip.

Astronautics: the science and technology of travel beyond Earth's atmosphere, including orbital motion and flight between the planets, moons, and eventually the stars.

Asymmetrical: a shape that has different features in opposite directions in relationship to a line or point.

Atmosphere: the outermost gaseous layer surrounding and bound to an object by gravity. For Earth this is primarily made up of nitrogen and oxygen.

Attitude: the current orientation in space in pitch, roll, and yaw.

Attitude Indicator (Artificial Horizon): a gyroscopic instrument that indicates the current position of the aircraft in both pitch and roll.

Autonomous: sensor- or timing-based operation with minimal human interference. Capable of making informed decisions based on sensory input.

Autonomous Mission: automatic execution of a task or set of tasks from beginning to end. Automated coordination between several systems to achieve a common task or goal.

Autonomous Vehicle System (AVS): a system that can operate, navigate, or perform tasks with little to no human intervention.

Autorotation: the technique for landing a rotorcraft without engine power. The primary method is to use the forces created by the descent to maintain a high speed for the rotors, and then, at the last possible moment, to greatly increase their lift to slow the rate of descent to zero just as the vehicle reaches the ground.

Avionics: the study of the electrical and mechanical instrumentation that indicates the status, condition, attitude, and future condition of a flying vehicle. Avionics includes controls for communications, navigation, engines, and aircraft movement.

B

Bernoulli's Principle: in a closed system, as the velocity of a fluid increases, the static pressure it creates decreases.

Binary Communication: Transmission of data digitally. The signal consists of values of 0 or 1, and the hardware or software receiving the transmission must decode the signal.

Bits: a binary digit represented by a 1 or 0; either an "on" state or "off" state.

Boundary Layer: a thin layer of air on the surface of an object in which the speed of the fluid flowing over the object changes from a relative velocity of zero at the surface to that of the freestream air.

C

Camber: the curvature of a surface. For an airfoil, the upper and lower camber describe the actual surfaces while the mean camber represents the geometric average interaction of the airfoil with the relative wind.

Cambered Airfoil: an airfoil shape that has a more pronounced curve on one half than the other; nonsymmetrical halves. In a relative wind a pressure difference between the two sides can result.

Celestial Sphere: the imaginary surface upon which we act as if all of the objects seen in the sky are located. The celestial sphere is a coordinate grid and is not an actual object.

Cellular Respiration: the reaction; every cell in the human body requires oxygen to release the energy that allows the cell to carry out its assigned function.

Center of Gravity (C.G.): every nut, bolt, and component is affected by gravity. The center of gravity is the point at which we can act as if the total force due to gravity was acting.

Center of Pressure: pressure varies from location to location on the wing or rotor of an aircraft. The center of pressure is the location at which we can act as if the total net pressure of aerodynamic forces is acting.

Ceramic: a material that is created by heating and cooling an inorganic, nonmetallic material. The material can contain crystals of the material or be glassy in composition.

Checklists and Standardized Procedures: a physical set of procedures that manages every critical aspect of an aircraft's or spacecraft's operations.

Chord Line: a straight line connecting the leading and trailing edges of an airfoil.

Circumlunar: a path that passes completely around the Moon.

Circumnavigation: a journey that goes around the entire planet.

Closed-Loop Control System: sensor-based system control. The system makes informed decisions based on outside input.

Coefficient: a term in a mathematical equation used to scale a numerical answer to the correct size. A number that is constant for a given substance, or process, serving as measure of a key characteristic.

Combustion: burning of a fuel; in an engine, it results in an expansion of the gas within the cylinder to push the piston downward.

Communication Protocol: description or rules for exchanging messages between systems. Protocols can be hardware or software based or any combination of the two.

Composites: a compound material that is formed out of two or more materials that remain distinct and separate components of a final structural piece.

Compression: created by forces that push inward on a material.

Computational Fluid Dynamics: the use of computer and algorithms to simulate and analyze the interactions between a moving fluid and the surfaces of a vehicle in order to predict the performance of the vehicle in real-world settings.

Control Surfaces: movable parts of an airplane designed to allow the pilot to control lift to maintain or change the motion of the airplane in pitch, roll, and yaw.

Corrosion: the weakening of material due to exposure to its environment as in the process of rusting or oxidation.

Cowl: a housing typically surrounding an engine that reduces drag while improving airflow around the engine.

Crankcase: the main body of an engine to which all other components are attached, may also be called the engine block.

Crankshaft: a complex shaft to which the pushrods from an engine's pistons are connected in order to transform their linear motion in the cylinder into the rotary motion of the shaft. The crankshaft can be connected to power an external device either directly or through a gear box to adjust the rate of the output shaft's rotation.

Cross-Sectional Area: the area represented by a slice through the object perpendicular to the direction of travel.

Cylinder: the housing in which an engine's pistons slide during the two or four strokes of the engine's power cycle. Cylinders contain the expanding combusting gases and transfer heat out of the engine.

D

Dampen: reduce the size of a vibration or repetitive motion.

Declination: the location of an object in the sky as measured north or south from the location of the celestial equator, a projection of Earth's equator out into space. Declination is typically measured in degrees.

Defense Advanced Research Projects Agency (DARPA): The research and development office for the U.S. Department of Defense. DARPA's mission is to maintain technological superiority of the U.S. military and prevent technological surprise from harming our national security. DARPA also creates technological surprise for our adversaries.

Deflection: the angle at which a control surface is displaced from its neutral position. A change in direction due to rebounding from a collision with a surface.

Digital Signals: data streams interpreted as 0 or 1.

Dihedral: a lifting of the wingtips to be higher than at the wing root. Dihedral can increase the roll (lateral) stability of an airplane.

Distance Measuring Equipment (DME): a specialized radio beacon that responds to an interrogation signal sent from the aircraft. Based on the time delay of the return signal, the instrument indicates a line-of-sight distance.

Downwash: the downward flow of air that is leaving the trailing edge of a wing as it moves through the air.

Drag: the resisting force produced by the relative motion of a vehicle through the air. Drag has two types. Parasitic drag is the result of frictional forces due to the flow of air around the vehicle. Induced drag is the result of the production of lift. In level flight, as a vehicle moves more quickly, the parasitic drag increases and the induced drag decreases.

Drag Coefficient: a single number that represents an airfoil shape's ability to transform its relative motion through the air into drag. The coefficient of drag is related to the actual shape of the airfoil and its angle of attack.

E

Electromechanical: a system that integrates both electrical and mechanical components to cause and control motion.

Elevator: a control surface attached to the trailing edge of a horizontal stabilizer. Deflection changes the angle of attack of the horizontal surfaces of the empennage creating aerodynamic forces that change the pitch of the aircraft's attitude.

Ellipse: the path followed by an object orbiting due to the force of gravity. A circle is a special case of an ellipse.

Empennage: the collective assembly at the tail of an aircraft. The empennage can include horizontal and vertical surfaces that are fixed or movable to control stability and maneuverability.

Energy: the total change that occurs, or can occur, in a system. Forms of energy include light, heat, gravitational potential, kinetic, chemical, nuclear, and electrical types. The total energy in a closed system is conserved.

Ephemeris Data: describes the precise location of any given satellite at every moment of time. This along with the almanac and pseudo-random code combine to allow the portable receiver to know exactly where, and when, each satellite's signal was transmitted.

Exhaust: 1) The flow of the products of combustion out of the cylinder and engine. 2) A stroke of a piston upward in a cylinder increases the pressure in the cylinder and pushes out the products of combustion.

Experimental Category: an airworthiness certification for an aircraft that has been constructed by amateur builders who have documented that they performed at least 51% of the assembly work to manufacture the vehicle.

Extravehicular Activity (EVA): operations completed by an astronaut outside of the protective interior of a spacecraft. EVAs are carried out in a spacesuit.

F

Fatigue: the weakening or breakage of a component due to repeated flexing.

Fibers: a slender filament of a material used in a composite to provide strength in one particular direction.

Flaps: control surfaces attached to the trailing edge of the wings on an aircraft. Flaps on both wings are deflected downward to increase the angle of attack of the wing and increase lift at lower airspeeds. Flaps can also create drag to slow the aircraft.

Flywheel: a large mass rotated by an engine to store its energy output for later use or to smooth out the power output curve of the engine.

Force: the push or pull exerted on an object.

Four Stroke: an internal combustion engine for which each piston travels the length of the cylinder a total of four times per combustion event. The strokes are typically intake, compression, power, and exhaust.

Fovea: region of high resolution, color vision on the retina.

Friction (Force): the resisting force that results from the relative or attempted motion of any two substances.

Fuselage: the complete central enclosure and framing to which an aircraft's wings, tail, and engines are attached.

G

Gauge Pressure: usually pressure measured above ambient atmospheric pressure.

G-forces: the perceived multiplier for weight. Maneuvering flight can change the multiplier to a negative value, zero (perceived weightlessness), or any positive value.

G-Induced Loss of Consciousness (G-LOC): the oxygen reserve in brain and retinal tissue in the eye runs out. Within seconds, the pilot transitions through the loss of color vision (grayout), tunnel vision, blackout and loss of consciousness.

Glide Ratio: the distance traveled forward divided by the altitude lost during the descent. The glide ratio is useful for predicting the range of an aircraft in a situation when engine power is lost such as during emergency or glider flight.

Glide: to fly without an internal source of energy such as an engine or motor.

Global Positioning System (GPS): a network of space satellites and receivers capable of accurately determining the location of the receiver anywhere and anytime on the surface of Earth.

Gravitational Potential: a form of energy that represents the future capability to fall to a lower elevation, thus it is based on the mass of the object, the local strength of gravity, and the height of the object.

Gravity: a field surrounding a mass that reaches to infinity but rapidly weakens with distance. The force that results from placing a second mass in the gravity field of another mass. The attractive force exerted equally but in opposite directions between any two masses.

Ground Effect: the increased lift and reduced drag experienced by an aircraft when it is operating within one wingspan of the ground (or surface).

G-Suit: a suit that contains air bladders that are hooked to the airplane and inflate nearly instantaneously when elevated G-forces are experienced in order to elevated the blood pressure in the lower extremities.

Gusset Plate: a reinforcing brace attached to one or more sides of a joint to strengthen the joint and brace its alignment.

Gyroscope (Gyroscopic Effect): a rapidly spinning mass has a tendency to maintain its orientation in space. This gyroscope can be used as a reference axis for determining the attitude of an aircraft.

Gyroscopic Control: the use of a rapidly rotating mass to create a stable axis that is the reference for steering a vehicle along a predetermined path.

H

Heading Indicator: a gyroscopic instrument that indicates the direction in which the vehicle is pointing its nose.

Heavier-Than-Air: a craft that has a total mass greater than that of the air it displaces. Heavier-than-air craft must create aerodynamic forces or have excess thrust that allow them to create lifting forces greater than the weight of the vehicle.

Helicopter: a heavier-than-air craft that produces its lifting force by rotating a set of rotors (wing like vanes) around a central point. Helicopters are capable of hovering in place for extended periods of time or climbing straight upward on the lift they produce.

Hohmann Transfer: a series of two rocket engine burns that initiate departure from a stable circular orbit around one object and entry into a stable circular orbit around a second object. An example is the transfer by the *Apollo* spacecraft as it traveled from Earth to the Moon.

Hybrid: combining two or more technologies or components to perform essentially the same task.

Hydrostatic Pressure: pressure exerted by a fluid (liquid or gas) at equilibrium due to gravity.

Hypersonic: a speed that is much larger than the speed of sound. Generally five or more times faster than Mach 1.

Hyperthermia: an unusually elevated body temperature.

Hypothermia: an unusually depressed body temperature.

Hypoxia: a reduced level of oxygen in the tissues of the body.

I

Inertia: an outcome of mass; all objects tend to continue moving in their current direction of travel unless acted upon by an outside force.

Innovation: a unique development or change in a product or service.

Instrument Landing System (ILS): a specialized radio beacon system that uses the difference in signal strength between two overlapping signals to indicate the aircraft's position in comparison to a flat plane in space. One plane is vertical and indicates the centerline of the approach, while another plane propagates upward from the threshold of the runway and indicates the glide slope to the touchdown area of the runway. Three vertical beams activate the outer marker indication for approach calibration, a middle marker for minimum decision altitude, and an inner marker that is about 1,000 feet from the runway threshold.

Instrument Rating: advanced certification.

Intake: 1) The flow of air and fuel, the reactants necessary for combustion, into the cylinder and engine. 2) A stroke of a piston downward in a cylinder, which lowers the pressure in the cylinder and draws air and fuel.

Intercontinental Ballistic Missile (ICBM): long-range ballistic missles that usually travel distances greater than 3,500 miles. Developed for nuclear warheads but also converted for initial space race.

ISR (Intelligence, Surveillance, Reconnaissance): the prime directive for most military UAVs and remote vehicles to gather information through observation and monitoring a situation.

J

Jig: a mechanical device that aligns individual parts for assembly into a larger component.

K

Kinetic Energy: the ability to cause change due to the speed of an object. Energy is stored as kinetic energy when the speed of the object increases and is released when the object slows down.

Kinetic: a form of energy that represents the capability to cause change due to an object's motion, thus it is based on the mass of the object and the square of its velocity times the constant 1/2.

L

Laminar Flow: the flow of a fluid in parallel layers in which each layer, or streamline, has a unique velocity, but there can be relative motion between layers.

Lateral Axis: a line through the vehicle's center of gravity that runs from wingtip to wingtip. Pitch is the rotational motion around the lateral axis. All three axes are at 90 degrees, or perpendicular, to each other.

Lateral Stability: stability in roll.

Leading Edge Slats: the first place on an airfoil that makes contact with the air through which it is moving.

Leading Edge: the first place on an airfoil that makes contact with the air through which it is moving.

Least Energy Mission: a specific flight path for a vehicle that requires the minimal amount of fuel to reach its intended location and flight direction and speed.

Lift Coefficient: a single number that represents an airfoil shape's ability to transform its relative motion through the air into lift. The coefficient of lift is related to the actual shape of the airfoil and its angle of attack.

Lift: in level flight, the upward force created by buoyancy and aerodynamic interaction with the air surrounding the aircraft. Lift forces can be varied and controlled to stabilize and maneuver the aircraft.

Linear Motion: motion in a straight line.

Line-of-Sight: unobstructed—by walls, earth, or other objects—pathway between two objects.

Liquid Fuel Rockets: rockets that use liquids for the fuel and oxidizer. Typical examples of liquid fuel and oxidizer are liquid hydrogen and oxygen.

Longitudinal Axis: in the direction of flight from nose to tail.

Longitudinal Wave: energy that travels parallel to the path or propagation of the wave energy such as compression or sound waves.

Lubrication: to reduce the amount of friction between two surfaces through the placement of another material between them such as oil, wax, or grease.

Lunar Roving Vehicle: compact vehicle sent to the lunar surface to greatly increase the area covered by lunar astronauts. Sent on Apollo 15, 16, and 17.

M

MAC Address: Media Access Control address. A unique number assigned to most network devices.

Mach Number (Mach Speed): a ratio between the actual speed of the airplane and the speed of sound under its current flight conditions. Thus, a mach number of 0.5 means the airplane is traveling at half the speed of sound, while a mach 3 flight is at three times the speed of sound.

Mass: a measure of the amount of material in an object. Mass is directly related to the inertia of an object and indirectly related through gravity to the weight of an object.

Materials Science: the study of the characteristics and properties of materials, shapes, structures, and their appropriate applications in science and engineering.

Maximum Lift to Drag Ratio or L/D Max: a balance point at which the least amount of total drag is produced.

Meridian: the line that runs from the South Pole to the North Pole through a location or a line of equal longitude on a map.

Microgravity: the experience of G-forces that are nearly zero due to the orbital or maneuvering flight of the aircraft or spacecraft.

Minimum Decision Altitude (MDA): the lowest altitude to which an aircraft can legally descend without making visual confirmation of the aircraft's position in relationship to a number of landing aids such as approach lighting, threshold lights, or runway lights, among others.

Moment: a measure of rotary forces. If there is a net moment, the object is not in equilibrium and will accelerate its rate of rotation.

Momentum: the quantity that describes the total motion of an object; the product of the object's mass and velocity.

N

Near-Earth Object (NEO): an asteroid or comet with an orbit that intersects or nearly intersects with that of planet Earth in terms of alignment in both space and time, thus an object with a significant likelihood of impacting Earth.

Newton's Third Law: for every force, there is an equal but opposite reaction force.

Non-Directional Beacons (NDB): a radio beacon that broadcasts a signal in all directions simultaneously.

Nozzle: a shaped channel that is designed to control and guide the exhaust gases from an engine.

O

Off-the-Shelf Components: standard materials or parts that can be found at a local hobby shop, hardware store, or other easily accessed vendor. Often parts are slightly modified from their original intent.

Open-Loop Control Systems: timing-based system with no outside input to control behavior.

Orbit: the curved path followed by an object as it moves around a planet or star or the act of moving along this curved path.

Orbital Elements: the numerical information required to accurately describe the orbital path of an object in both space and time.

P

Perigee: the point of nearest approach between an orbiting body and Earth.

Perihelion: the point of nearest approach between an orbiting body and the Sun.

Period: the time required to complete one complete orbit.

Piston: the movable wall of the combustion chamber. The rapidly expanding combustion products push the piston downward in the cylinder. This motion is transferred to the crankshaft and propeller through piston rods with the continuing rotation of the crankshaft forcing the piston back upward in the cylinder.

Pitch: 1. the distance that a propellor or fan would advance during a single rotation if there was no slipping between it and the fluid through which it is rotating. 2. the rotational movement of the vehicle around the lateral axis such that the nose of the vehicle moves up and down. Pitch is generally stabilized by the horizontal stabilizer and controlled with the elevator.

Pitch: the distance that a propeller or fan advances during a single rotation if there is no slipping between it and the fluid through which it is rotating.

Pitch Stability: depends heavily on weight, balance, and the angle at which the horizontal stabilizer is mounted on the fuselage. The horizontal stabilizer on most aircraft is mounted with a slightly negative angle of attack.

Planform: the shape of the wing as it appears in a top view of the aircraft.

Plugs-Out Test: the procedure of completely sealing a spacecraft on the ground to test and evaluate the performance of all internal systems.

Potential Energy: the ability to cause change due to the position of an object and its ability to fall at some time in the future.

Powerplant: the combination of engine and propeller, turbines, or fan that transform energy from some source (chemical, electrical, mechanical) into thrust.

Power-to-Weight Ratio: an expression of the efficiency for an engine or vehicle by comparing the amount of energy that it can release per unit time to its total weight.

Precession: the change in alignment between the orbital plane and the tilt of an object's rotational axis over time due to torque. The same term applies to the wobble of a toy top.

Propulsion: the system or engine used to transform energy into a mechanical force. Propulsion forces are used to overcome drag and to stabilize and maneuver the craft.

Pseudo-Random Code: in a GPS, this repeating sequence of nonrandom numbers is broadcast at very low power levels. When the signal is received by a small antenna and integrated over time, a positive identification of the signal can be made with very low-cost receiving equipment.

R

Radioisotope Heaters: a device that uses the heat released by the radioactive decay of a material as a source of thermal energy.

Ramjet: an engine with no moving parts that relies on its rapid forward motion and internal shape to compress the air in the atmosphere before combining it with fuel and heat for combustion. A ramjet's airflow occurs at supersonic speeds; however, the air slows to subsonic as it is compressed.

Reaction Control System (RCS): a means to maneuver a vehicle by exerting action forces to spray material away from a vehicle with the reaction force on the vehicle pushing it in the opposite direction of the exhaust jet.

Reinforcement: a material incorporated into a structure to add strength.

Remote Piloted Vehicle (RPV): vehicle where the pilot or operator is not physically in the vehicle. Control is through a means of wired or wireless interaction.

Resin: a material used to bind the fibers of a composite material in place.

Reynolds Number: an indication of the ratio of the inertial and viscous forces in a flow. Of critical importance is that any aerodynamic test has to be performed at the same Reynolds number as that of the real environment. Thus, a half-scale model must be tested at twice the real-world velocity.

Rib: the physical structure that holds the skin of a wing in the proper aerodynamic shape.

Right Ascension (RA): the location of an object in the sky measured eastward from the location of the vernal equinox (location of the Sun at the first moment of spring). Right ascension is typically measured in degrees or hours.

Rivet: A mechanical fastener formed as a cylinder with a head on one end. A permanent mechanical connection is made after the rivet is placed through holes in multiple layers of material, and then the cylindrical end is deformed to form a secondary head.

Rocker-Boogie: a component of a drive train system that uses multiple wheels on a single support mechanism that can flex and move relative to the chassis of the vehicle.

Rocket (Liquid Fuel): a reaction engine in which the original thrust force is created by combusting and exhausting liquid fuel and oxidizer.

Rocket (Solid Fuel): a reaction engine in which the original thrust force is created by combusting and exhausting solid fuel and oxidizer.

Rocket Engine: a reaction engine in which the original thrust force is created by combusting and exhausting fuel and oxidizer, which are entirely derived from materials carried onboard the vehicle.

Roll: the rotational movement of the vehicle around the longitudinal axis. Roll is generally stabilized by the dihedral and position of the wing and changed by movement of the ailerons.

Rotational Motion (Rotation): movement around the center of mass.

Rudder: the movable vertical surface mounted to the vertical stabilizer. The rudder deflects to create aerodynamic forces that push the tail to the left or right. The aerodynamic forces cause rotation about the vertical axis, or yaw.

S

Scramjet: an engine with no moving parts that relies on its rapid forward motion and internal shape to compress the air in the atmosphere before combining it with fuel and heat for combustion. A scramjet's airflow occurs at supersonic speeds with the air flow through the engine remaining supersonic as it is compressed, combusted, and exhausted.

Skin: the material in aerodynamic contact with the relative wind or external environment of the vehicle.

Smart Metal: after this material is formed, it can alter its specific shape while remaining a solid based on changes in temperature, electrical currents, or other energy inputs.

Solar Arrays: an assembly of photovoltaic devices that gather light to generate electrical energy.

Solid Fuel Rockets: rockets that use paste or thickened slurry composition for a fuel and oxidizer source. The solid fuel is usually prepacked or stacked in sections and, once ignited, can be difficult to control.

Solid State Control: a system made out of components that contain the flow of electricity entirely within solid components in contrast to vacuum tube or gas-filled tubes.

Solid State: most modern devices are solid-state devices wherein the electrons are contained within solid materials, not gas tubes or relays. Commonly associated with printed circuit boards (PCB) and increasing miniaturization of the technology.

Somatogravic Illusion: the pilot perceives that the aircraft is pitching violently upward.

Sound Barrier: the dramatic increase in drag as a vehicle approaches and reaches the speed of sound represent a true barrier that requires additional power or drag reduction techniques to exceed the speed of sound and go supersonic.

Space Station: an orbiting assembly of habitats, power, communication, life support, and research devices to support human exploration of space near a planet.

Spar: the main structural element of a wing's framework. The spar is designed to stiffen the structure and transfer forces back to the fuselage of the vehicle.

Spatial Awareness: allows us to smoothly walk, run, jump, and otherwise maintain our balance.

Spatial Disorientation: an inaccurate perception of the attitude and motion of the vehicle.

Spoilers: mechanical devices designed to extend into the airflow over an airfoil in order to create turbulence, reduce lift, and induce roll.

Sputnik: the first human crafted artifact to be placed into orbit around Earth on October 4, 1957.

Stability: a description of how an object responds to being disturbed from its initial position and motion.

Stall: beyond the critical angle of attack (AOA), an airfoil rapidly reduces the amount of lift that it produces. This is a stalled condition.

Stator: a multibladed, nonrotating section of gas turbine engine that straightens the fluid flow along the flow axis of the engine so that more energy can be transferred into or out of the gases.

Supersonic: to move at a speed that is faster than the speed of sound.

Supervised Autonomy: a type of remote system where a human operator can tell the vehicle where it should go, but the sensors on the vehicle determine the best way to get there.

Sweepback: the angle at which the wing's leading edge is mounted to the fuselage compared to the lateral axis.

Symmetrical: having an identical but opposite shape across a line or plane of symmetry.

Synthetic Vision: a system of instrumentation and displays that create an accurate and timely environment in the cockpit that allows visual flight procedures to be used for navigation.

Systems Engineering: the discipline of engineering that deals with identifying system goals, designs, constraints, feature importance, tradeoffs, risk management, and ultimately the best design for the customer and task needs.

Systems of Systems (SoS): multiple interconnected but independent systems working together to form a large more valuable system.

T

Telemetry Data: the measurement and reporting of data allows a remote system operator to analyze data on the fly.

Test Section: the volume of air in a wind tunnel in which the air has already been prepared by being brought up to a uniform speed and laminar flow in one direction so that an object placed in this volume can be evaluated for its aerodynamic performance.

Thrust: the force produced to overcome drag and momentum to control the velocity and position of the aircraft.

Time of Useful Consciousness: reduction in air pressure.

Torque: a force applied through any point other than the center of mass of an object. A force that induces a rotational motion. Mathematically, the product of a force and the distance at which it is applied from the center of rotation.

Trailing Edge: the last place on an airfoil that makes contact with the air through which it is moving.

Transonic Drag Rise: the rapid increase in drag that occurs as an aircraft gets very close to the speed of sound. The drag appears as though it will rise indefinitely but reaches a maximum at some speed after which it rapidly decreases again.

Transverse Wave: energy that travels perpendicular to the path or propagation of the wave energy such as a ripple on water or the pluck on a musical string instrument.

Triaxial: consisting of movement or control along the x-, y-, and z-axis of motion in three-dimensional space.

Trim Tab: a small movable control surface attached to a larger control surface such as the rudder, elevator, or ailerons so that it can create an aerodynamic force to hold the larger control surface in a given position without any control pressure in the cockpit.

Tsiolkovsky Equation: a mathematical tool that relates the change in velocity a rocket will experience (delta-v) based on its effective exhaust velocity and the initial and final masses of the rocket.

Turbine: a multibladed design optimized to capture the energy from a flowing fluid and transform it into rotary motion. Typically used in a gas turbine (jet) engine to capture the energy required to run the compressor stage of the engine as well as auxiliary devices.

Turbocharger: a device that compresses the air entering an engine so that it can combust more fuel per unit time, thus increasing the power output of the engine.

Turbofan: a jet engine in which the shaft power rotates a fan blade assembly that is larger in diameter than that of the compressor blades so that there is a large volume of unheated bypass air created. This increases the thrust produced by the engine at slower flight speeds than those for a turbojet engine.

Turbojet: a jet engine in which the exhaust gases from the engine directly provide the thrust. This is most efficient for providing thrust at very high subsonic, transonic, and supersonic speeds.

Turboprop: a jet engine in which the shaft power is put through a gear reduction system to turn a propeller so that a very large volume of unheated bypass air is created. Turboprops benefit from the continuous flow rate of a jet engine and the efficiency of propeller-driven flight for slower flight speeds.

Turbulent Flow: the flow of a fluid in which the velocity in the flow is nonuniform and variable at any given point.

Turbulent Flow: the flow of a fluid such that the velocity in the flow is nonuniform and variable at any given point.

Turn Coordinator: a gyroscopic instrument that indicates whether the aircraft is in a slip, skid, or coordinated flight.

Two-Stroke: an internal combustion engine for which each piston travels the length of the cylinder a total of two times per combustion event. The strokes are compression (with intake during its beginning moments) and combustion (with exhaust at its beginning moments).

U

Upwash: the zone in front of a wing in which the airflow just in front of the approaching airfoil is deflected upward.

UV (Ultraviolet): an intense form of radiant energy (light) in sunlight that can cause damage to materials such as fabric, wood, plastic, and living tissue.

V

Valves: a device that seals either an intake or exhaust port on a cylinder except at the times required for the flow of the gases into or out of the cylinder.

Vector: an arrow that represents both the direction and magnitude of a value by its orientation in space and the length of the arrow.

Vernier Thrusters: a small rocket motor used to create forces that finely tune the attitude of the spacecraft in conjunction with a larger spacecraft propulsion system.

Vertical Axis: perpendicular to the other two axes.

Vertical Speed Indicator (VSI): a Pitot-Static instrument that indicates the rate of climb or descent.

Very High Frequency, Omnidirectional Radio Beacon (VOR): a navigational aid that uses the phase shift between an omnidirectional and rotating beacon to create individual 1° degree wide paths that radiate from the transmitter.

Vortex(Vortices): a swirling or circular motion in a fluid or gas.

W

Wave Drag: the significant increase in drag produced due to the creation of transonic pressure fronts as a vehicle approaches and exceeds the speed of sound.

Weight Shift Aviation: a means of controling an aircraft in flight by moving weight, typically the pilot, to change the orientation of the aircraft and the resulting aerodynamic forces.

Weight Shift: a means of controlling an aircraft in flight by moving weight, typically the pilot, to change the orientation of the aircraft and the resulting aerodynamic forces.

Weight: the downward force on an object due to gravity.

Weld: binding two materials together by temporarily melting both materials so that they become one when they solidify.

Wing Warping: the twisting of the entire wing panel in response to a control input to change the angle of attack at the wingtip. Wing warping is used to change the lift produced by each wing panel and induce roll.

Y

Yaw: the rotational movement of the vehicle around the vertical axis. Yaw is generally stabilized by the vertical stabilizer and changed by movement of the rudder.

Z

Zenith: the point in the sky that is directly above your current location.

INDEX